Lecture Notes in Mathematics 1595

Robert D.M. Accola

# Topics in the Theory of Riemann Surfaces

Springer-Verlag
Berlin Heidelberg New York
London Paris Tokyo
Hong Kong Barcelona
Budapest

Author

Robert D. M. Accola
Department of Mathematics
Brown University
Providence, Rhode Island 02912, USA

Mathematics Subject Classification (1991): 30F10; 14H15

ISBN 3-540-58721-7 Springer-Verlag Berlin Heidelberg New York

CIP-Data applied for



Typesetting: Camera-ready out put by the author
SPIN: 10130221 46/3140-543210 - Printed on acid-free paper

## Preface

These are lecture notes for a course given during the Fall of 1988 at Brown University. The students were assumed to have had a previous course on the theory of Riemann surfaces or algebraic curves. Chapter One of these notes gives a review of most of the basic material needed later, but few proofs are given, and the reader is assumed to have some previous acquaintance with the material.

Besides giving the author's point of view and notational conventions, the introduction gives a few proofs (or rather demonstrations) intended to bridge the gap between the more analytic approach to the subject as exemplified by the books of Ahlfors-Sario [4], Farkas-Kra [11] or Forster [12] and the more algebraic approach as exemplified by Walker [26]. In particular the proof of the genus formulas for a plane curve based on the Riemann-Hurwitz formula and other materials in Section 1.3 were inspired by a seminar given in the early '60's at Brown by S. Lefschetz.

Despite the recent appearance of several excellent books on the theory of compact Riemann surfaces (which is, of course, basically the same as the theory of algebraic curves over the complex numbers), most of the material in these notes does not appear in these books. The two main subjects treated here are exceptional points on Riemann surfaces (Weierstrass points, higher-order Weierstrass points) and automorphisms of Riemann surfaces. A foundational treatment of the theory of automorphisms from the viewpoint of Galois coverings of Riemann surfaces is given in Chapters Four and Five, following and expanding to some extent the treatment of Ahlfors-Sario [4] and Seifert-Threlfall [25]. The treatment is technically different from that of A. M. Macbeath [19] and his students, although fundamentally it is the same. In the treatment here, no mention of Fuchsian groups occurs. Also a treatment of the extremely useful inequality of Castelnuovo-Severi is included, a treatment for which the author knows no reference. (It is difficult, however, to believe that anything on this venerable subject can be really new.)

The treatment of all subjects is basically elementary.

The author heartily thanks Natalie Ruth Johnson for an excellent job in preparing this manuscript.

## Table of Contents

## Chapter 1. Review of some basic concepts in the theory of Riemann surfaces.

1.1 Coverings. A surface $X$ is a connected Hausdorff space which satisfies the second axion of countability and has a basis for the open sets of sets homeomorphic to open sets in $\mathbb{C}$, the complex numbers. A Riemann surface is a surface with an open cover $\{U_\alpha \mid \alpha \in A\}$, for some index set $A$, and homeomorphisms $\{\varphi_\alpha\}$, $\varphi_\alpha : U_\alpha \to \mathbb{C}$, where $\varphi_\beta \circ \varphi^{-1}_\alpha$ is biholomorphic wherever it is defined. A pair $(U_\alpha, \varphi_\alpha)$ is called a chart and the set $\{(U_\alpha, \varphi_\alpha) \mid \alpha \in A\}$ is called an atlas. $\varphi_\alpha$ will be called a local parameter. All Riemann surfaces are orientable.

A continuous function $f : X \to Y$ between Rieman surfaces will be said to be holomorphic if it is holomorphic (analytic) when expressed in local parameters. If $f$ is holomorphic, injective, and surjective then $f$ is said to be biholomorphic and $X$ and $Y$ are said to be conformally equivalent. A biholomorphic map of a Riemann surface onto itself will be called an automorphism, and the group (under composition) of automorphisms of a Riemann surface $X$ will be denoted $\mathrm{Aut}(X)$. An automorphism of period 2 will be called an involution.

A holomorphic (meromorphic) function on a Riemann surface $X$ is a holomorphic mapping of $X$ into $\mathbb{C}$ ($\mathbb{P}^1$, the projective line, or Riemann sphere).

If $f : X \to Y$ is a non-constant holomorphic mapping of Riemann surfaces and $x \in X$, then in suitable local coordinates at $x$ and $f(x)$, $f$ looks like $z \to z^n$, $n$ a positive integer. We say $f$ is equivalent to $z \to z^n$ at $x$. (In particular, $f$ is an open mapping.) $n$ is said to be the multiplicity of $f$ at $x$, and $n - 1$ is said to be the branching or ramification of $f$ at $x$, denoted $\mathrm{ram}_x(f)$. The ramified points $x$, where $\mathrm{ram}_x(f) > 0$, are a discrete set in $X$.

Theorem. Let $f : X \to Y$ be a non-constant holomorphic map between compact Riemann surfaces. Then there exists a positive integer $n$ so that for any $y \in Y$

$$n = \mathrm{card}\left\{f^{-1}(y)\right\} + \sum_{f(x)=y} \mathrm{ram}_x(f) .$$

$n$ will be called the number of sheets in the covering $X \to Y$. If $Y = \mathbb{P}^1$ then $f$ is a non-constant meromorphic function on $X$ and

$n$ is called the order of $f$, denoted $o(f)$.

A non-constant holomorphic map of compact Riemann surfaces will often be called a covering.

The Riemann-Hurwitz Formula for coverings.

Let $f : X \to Y$ be a non-constant $n$-sheeted holomorphic map between compact Riemann surfaces. Let $p$ and $q$ be the genus of $X$ and $Y$ respectively. Let $\mathrm{ram}(f) = \sum_{x \in X} \mathrm{ram}_x(f)$. Then

$$2p - 2 = n(2q - 2) + \mathrm{ram}(f).$$

Definition. If $X_p$ is a compact Riemann surface, the subscript $p$ will always denote the genus of $X_p$.

Corollary. Let $f : X_p \to X_q$ be a non-constant $n$-sheeted holomorphic mapping of compact Riemann surfaces where $n \geq 2$. Then $p \geq q$ with equality possible only if $p = 0$ or $1$.

1.2 Function Fields. If $X$ is a Riemann surface let $M(X)$ denote the field of meromorphic functions on $X$. If $f : X \to Y$ is an $n$-sheeted holomorphic map then $f^* : M(Y) \to M(X)$ maps $M(Y)$ onto a field which is of index $n$ in $M(X)$. Conversely, if $X$ is a compact Riemann surface and $K$ is a subfield of $M(X)$ of index $n$, then there exists an $n$-sheeted covering $f : X \to Y$ of compact Riemann surfaces and $f^*(M(Y)) = K$.

A rational function field is a field isomorphic to $M(\mathbb{P}^1)$, that is, the rational functions. A meromorphic function field on a Riemann surface of genus 1 wil be called an elliptic function field. A Riemann surface $X_p$, $p \geq 2$, will be called hyperelliptic if $M(X_p)$ admits a rational subfield of index 2. This is equivalent to $X_p$ admitting a 2-sheeted covering $X_p \to \mathbb{P}^1$. The interchange of the two sheets of this covering is an automorphism of $X_p$ of order 2 called the hyperelliptic involution. A hyperelliptic Riemann surface, $X_p$, is the Riemann surface for a polynomial $P(z,w) \in \mathbb{C}[z,w]$ where

$$P(z,w) = w^2 - \prod_{j=1}^{2p+2} (z - a_j) \text{ where } a_i \neq a_j.$$

A Riemann surface of genus greater than one will be called elliptic-hyperelliptic if it admits a 2-sheeted covering of a Riemann surface of genus one ; that is, its function field admits an elliptic function field of index 2 .

In general, if $X$ is a compact Riemann surface then $M(X)$ is the Riemann surface for some (many) irreducible polynomial $P(z,w) \in \mathbb{C}[z,w]$ .

1.3 Plane Curves [26]. Let $f(x,y,z)$ be an irreducible homogeneous polynomial of degree $n$ . Let $C_n = \{(x,y,z) \in \mathbb{P}^2(\mathbb{C}) \mid f(x,y,x) = 0\}$, an irreducible plane curve. Suppose $C_n$ is non-singular, that is, $f_x(x,y,z) = 0 = f_y(x,y,z) = f_z(x,y,z)$ implies $(x,y,z) = (0,0,0)$ .

We show that as a Riemann surface, $C_n$ has genus $\frac{(n-1)(n-2)}{2}$ . $C_n$ has only a finite number of inflection points (where the tangent line has more than two intersections with $C_n$ ). Choose coordinate axes in $\mathbb{P}^2$ so that none of the inflectional tangent lines pass through $(0,1,0)$ (i.e. no inflectional tangent line is parallel to the y-axis). Then each tangent line to $C_n$ passing through $(0,1,0)$ has only two intersections where it is tangent. Such points occur where $f = 0$ and $f_y = 0$ . At such points the projection of $C_n$ onto the x-axis is locally two-to-one; that is, this projection (a holomorphic map onto the x-axis ($= \mathbb{P}^1$)) has ramification one at each of these points. The number of such points is $n(n-1)$ .

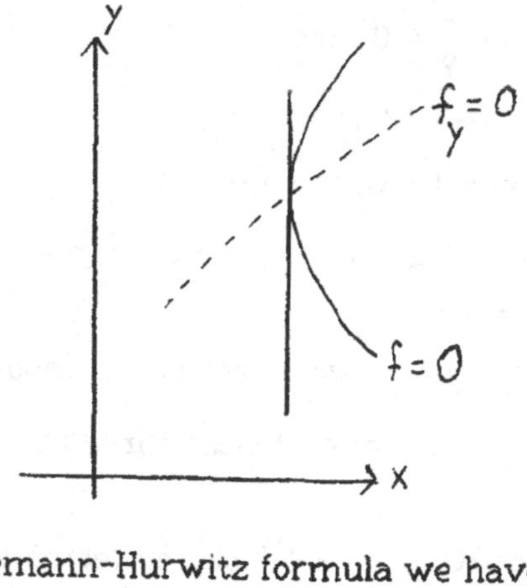


Using the Riemann-Hurwitz formula we have

$$2p - 2 = -2n + n(n-1) \ ; \ 2p - 2 = n(n-3)$$

$$\text{or} \quad p = \frac{n^2 - 3n + 2}{2} .$$

We can use the same method if $C_n$ has singularities. Suppose $C_n$ has an ordinary singularity of multiplicity $k$ at $(0,0,1)$. Dehomogenizing $f(x,y,z)$ to $f(x,y,1)$ we have

$$f = \prod_{i=1}^{k} (\alpha_i x + \beta_i y) + \text{higher order terms}$$

where the $k$ tangents lines $\alpha_i x + \beta_i y = 0$ are all distinct. Thus

$$f_y = \prod_{i=1}^{k-1} (\gamma_i x + \delta_i y) + \text{higher order and so } f = 0 \text{ and } f_y = 0 \text{ have}$$

$k(k-1)$ intersections at $(0,0,1)$. Now assume $C_n$ has $s$ ordinary singularities of multiplicities $k_1,\ldots,k_s$. Then the total ramification

for the projection of $C_n$ onto the $x$-axis is $n(n-1) - \sum k_j(k_j - 1)$ or

$$p = \frac{(n-1)(n-2)}{2} - \sum \frac{k_j \cdot (k_j - 1)}{2}.$$

Let us look more closely at singularities of multiplicity two at, say, $(0,0,1)$.

$f = ax^2 + bxy + cy^2 +$ higher order.
$f_y = bx + 2cy +$ higher order.

Local parametric equations for $f_y = 0$ are

$$x = 2ct + c_2 t^2 + \cdots$$

$$y = -bt + b_2 t^2 + \cdots .$$

$$\text{Thus } f(x,y) = a(4c^2t^2 + \cdots) + b(-2bct^2 + \cdots) + c(b^2t^2 + \cdots) + o(t^3)$$
$$= c(4ac - b^2)t^2 + o(t^3) .$$

If $4ac - b^2 \neq 0$, $f = 0$ and $f_y$ have two intersections (node).

If $4ac - b^2 = 0$, $f = 0$ and $f_y = 0$ have at least three intersections (cusp).

If $f = (ax + by)^2 + (ax + by)Q(x,y) + 4^{th}$ degree terms and higher order, then $f = 0$ and $f_y = 0$ have 4 intersections (tacnode).

Nodes add nothing to the ramification of the propection map. Cusps add one to this ramification and tacnodes add nothing.

If $\delta$ = # of nodes, $\kappa$ = # of simple cusps, $\eta$ = # of simple tacnodes for $C_n$ and there are no further singularities then

$$2p - 2 = -2n + n(n-1) - 2\delta - 3\kappa - 4\eta + \kappa$$

$$p = \frac{(n-1)(n-2)}{2} - \delta - \kappa - 2\eta .$$

For a general plane curve of degree $n$ we have

$$p = \frac{(n-1)(n-2)}{2} - \delta$$

where $\delta$ will be called the $\delta$-invariant (sometimes called the number of nodes suitable counted).

Ordinary k-fold singularities add $\frac{k(k-1)}{2}$ to the $\delta$-invariant.
Simple cusps add 1 to the $\delta$-invariant.
Simple tacnodes add 2 to the $\delta$-invariant.

1.4 Divisors on Riemann surfaces. A divisor $D$ on a Riemann surface $X$ is a singular zero-chain, written $D = \sum n_i z_i$ for $n_i \in Z$ and $z_i \in X$. The degree of $D$, deg $D$, is $\sum n_i$, an integer. If $n_i \geq 0$ for all $i$ then $D$ is said to be positive (or integral) and we write $D \geq 0$. The zero divisor is positive.

If $D = \sum n_i z_i$ and $D' = \sum n'_i z_i$ are two divisors then by the greatest common divisor of $D$ and $D'$, written $(D, D')$, we mean the divisor $\sum \min(n_i, n'_i) z_i$. Thus to write $(E, z) = 0$ is to say that the coefficient of $z$ in $E$ is zero. If $(E, F) = E$ we will write $E \subseteq F$.

If $f$ is a meromorphic function on $X$, the divisor of $f$ is written $(f)$ where $(f)$ = (zeros of $f$) - (poles of $f$)

$$= (f)_0 - (f)_p .$$

Note that $\deg(f) = 0$.

Two divisors $D, E$ are said to be (linearly) equivalent if there exist an $f \in M(X)$ and $(f) = D - E$ (an equivalence relation).

Notation: $D \equiv E$.

Definition. If $D$ is a divisor, $L(D) = \{ f \in M(X) \mid (f) + D \geq 0 \}$.

Theorem. If $X$ is compact then $L(D)$ is a finite dimensional vector space over $\mathbb{C}$.

Definition. If $D$ a divisor on $X$, $\ell(D) = \dim_{\mathbb{C}} L(D)$, the affine dimension of $L(D)$. $r(D) = \ell(D) - 1$, the projective dimension of $L(D)$.

A meromorphic (or abelian) differential $\omega$ is a one-form that locally can be written

$$\omega = f(z)\,dz$$

where $f$ is meromorphic. The value of a meromorphic differential at a point of $X$ is not well defined; however, zeros and poles are well defined so we can consider the divisor of $\omega$,

$$(\omega) = (\text{zeros of } \omega) - (\text{poles of } \omega).$$

If $\omega_1$ and $\omega_2$ are two meromorphic differentials then $\omega_1/\omega_2$ is a well defined meromorphic function on $X$. Since $\deg(\omega_1/\omega_2) = 0$ we see that $\deg(\omega_1) = \deg(\omega_2)$.

Theorem. If $\omega$ is a meromorphic differential on $X_p$ then

$$\deg(\omega) = 2p - 2.$$

Definition. If $D$ is a divisor

$$\Omega(D) = \{\text{meromorphic differentials } \omega \,|\, (\omega) - D \geq 0\}$$

$i(D)$ = dimension of $\Omega(D)$ as a vector space over $\mathbb{C}$.

Lemma. If $K$ is the divisor of a meromorphic differential then $i(D) = \ell(K - D)$.

Now we list some of the basic theorems concerning compact Riemann surfaces.

Riemann-Roch theorem. If $D$ is a divisor on a compact Riemann surface of genus $p$, then

$$r(D) = \deg D - p + i(D).$$

Brill-Noether formulation of the Riemann-Roch theorem. If $D$ and $D'$ are two divisors so that $D + D' \equiv (\omega)$ where $\omega$ is a meromorphic differential, then $\deg D - 2r(D) = \deg D' - 2r(D')$.

Notice that the divisors in the Brill-Noether theorem need not be positive.

Definition. A meromorphic differential without poles is called holomorphic (or analytic, or regular, or an abelian differential of the first kind).

Definition. If $D \geq 0$ and $i(D) > 0$ then $D$ is said to be special and $i(D)$ is often called the index (of speciality) of $D$.

Clifford's theorem. If $D$ is special then $\deg D - 2r(D) \geq 0$.

Definition. If $D$ is a divisor, the Clifford index of $D$, denoted $c(D)$, is:

$$c(D) = \deg D - 2r(D) = p - r(D) - i(D).$$

Strong form of Clifford's theorem. If $X_p$ admits a special divisor $D$ where $\deg D \neq 0$, $\deg D \neq 2p-2$ and $c(D) = 0$, then $X_p$ is hyperelliptic.

Definition. $\Omega(X_p) = \{$holomorphic differentials on $X_p\}$.

Corollary. $\dim_{\mathbb{C}} \Omega(X_p) = p$. If $\omega \in \Omega(X_p)$ then $i\big((\omega)\big) = 1$.

1.5 Linear series. Let $D$ be a divisor on $X_p$. Then the complete linear series determinant by $D$, denoted $|D|$, is

$$|D| = \left\{D' \geq 0 \,\middle|\, D' \equiv D\right\}.$$

$|D|$ can be empty. If $D \geq 0$ then $|D|$ is a set of divisors parametrized by the projectification of $L(D)$.

$$|D| = \left\{D' \,\middle|\, D' - D = (f),\ f \in L(D)\right\}.$$

Let $S$ be a linear subspace of $L(D)$ of dimension $r+1$, $r \leq r(D)$. The set of divisors parametried by $S$

$$\left\{D' \,\middle|\, D' - D = (f),\ f \in S\right\}$$

is called a linear series and denoted $g^r_n$ where $n = \deg$ and $r+1 = \dim S$.

If $r < r(D)$, $g^r_n$ is said to be incomplete.

Example: $K$ = divisors of holomorphic differentials = $g^{p-1}_{2p-2}$. $K$ is the unique linear series of dimension $p-1$ and degree $2p-2$.

A linear series $g^r_n$ can have base (or fixed) points, that is, a divisor common to all divisors in $g^r_n$. If $F$ of degree $f$ is such a divisor then $g^r_n - F = g^r_{n-f}$ stands for the linear series without base points obtained by subtracting $F$ from each divisor in $g^r_n$.

Let $g^r_n$ be a linear series without base points. Then we can map $X$ into $\mathbb{P}^r$ as follows. Suppose

$$g^r_n = \left\{D' \,\middle|\, D' - D = (f),\ f \in S\right\},\ S \subset L(D).$$

Let $f_0, \cdots, f_r$ be a basis for $S$. If $x \in X$ let

$$\Theta(x) = \big(f_0(x), \cdots, f_r(x)\big) \in \mathbb{P}^r(\mathbb{C}).$$

If we pick a local parameter $z$ at $x_0$, $z(x_0) = 0$ then

$$f_i = z^{n_i} g_i(z) \text{ where } g_i(0) \neq 0 .$$

Locally $z$ is mapped by $\Theta$ into $\left(\cdots, z^{n_i} g_i(z), \cdots\right) \in \mathbb{P}^r$.

Let $n = \min\{n_i\}$. Then $z \overset{\Theta}{\mapsto} \left(\cdots, z^{n_i - n} g_i(z), \cdots\right)$ and so

$x \overset{\Theta}{\mapsto} \left(\cdots, z^{n_i - n} g_i(z)\Big|_{z=0}, \cdots\right)$ and at least one component is non-zero. That is, $\Theta$ is well defined on all of $X$.

Let $C = \Theta(X) \subset \mathbb{P}^r$ be the image of $X$ under $\Theta$. Then the hyperplane sections of $C$ pull back via $\Theta$ to the divisors of $g^r_n$. For

if $\sum_{i=0}^{r} a_i y_i = 0$ is a hyperplane $H$ in $\mathbb{P}^r$ then

$\Theta(x) \in H \cap C \Leftrightarrow \sum a_i f_i(x) = 0 \Leftrightarrow x$ is a zero of $\sum a_i f_i$. If $f_r \equiv 1$ then $D$ is the pull back of the hyperphane (at $\infty$) $y_r = 0$.

If the map $\Theta$ is one-to-one in general, then $g^r_n$ is said to be <u>simple</u>. In this case $X$ is (conformally equivalent to) the Riemann surface (or normalization) of $C$. ($C$ can, of course, have singularities.) Each hyperplane in $\mathbb{P}^r$ cuts $C$ in $n$ points (counting multiplicities) so we write $C_n$ for $C$, a curve of degree $n$ in $\mathbb{P}^r$.

If the map $\Theta$ is not one-to-one in general then $g^r_n$ is said to be <u>composite</u>. Then $X$ is a t-sheeted covering $(t \geq 2)$ of (the normalization of) $C$ and $C$ has degree $n/t$, since $g^r_n$ has no base points. (Thus a linear series of dimension $r \geq 2$ of prime degree greater than $r$ without base points is simple.) In this case, if $Y_q$ is the normalization of $C$ then $Y_q$ admits a $g^r_{n/t}$. If $\pi : X_p \to Y_q$ is the t-sheeted covering then $g^r_{n/t}$ on $Y_q$ lifts to $g^r_n$ on $X_p$. The fibers of this map $\pi$ are called an <u>involution of genus $q$</u>

and denoted $\gamma^1_n$. In this case we say that $g^r_n$ is compounded of the involution $\gamma^1_t$. (In the sequel, we shall be careful not to confuse the two meanings of the word "involution".)

Given $g^r_n$ and $\Theta : X \to \mathbb{P}^r$, $\Theta$ depends on the choice of bases $\{f_0, \cdots, f_r\}$ in $S \subset L(D)$. A second choice of basis gives another curve in $\mathbb{P}^r$, but the two curves are related by a projective transformation whose matrix is that which relates the two bases of $S$.

If $g^2_n$ is simple, then $\Theta(X) \subset \mathbb{P}^2$ is called a plane model for $X$.

For a $g^1_n$ without base points $\Theta : X \to \mathbb{P}^1$ corresponds to an n-sheeted covering of $\mathbb{P}^1$. If $X_p$ $(p \geq 2)$ admits a $g^1_2$ then $X_p$ is hyperelliptic, and conversely.

If $X_p$ admits a $g^r_r$ then $p = 0$. If $X$ admits a complete $g^r_{r+1}$, $r \geq 2$, then $p = 1$.

The canonical curve is the image of $X_p$ by $\Theta$ in $\mathbb{P}^{p-1}$ given by the canonical series $g^{p-1}_{2p-2}$. $g^{p-1}_{2p-2}$ is base point free. It is composite if and only if $X_p$ is hyperelliptic in which case it is compounded of the $g^1_2$.

For a non-hyperelliptic $X_3$, the canonical curve is a non-singular plane quartic, $C_4$ where the lines in $\mathbb{P}^2$ cut out the canonical series.

If $g^r_n$ is a complete linear series and $D \in g^r_n$ then $r = n - p + i(D)$. Since $i(D)$ does not depend on the divisor chosen, we will occasionally write $i(g^r_n)$ in place of $i(D)$.

If $g^r_n$ is a complete linear series and $D$ is an integral divisor, by $|g^r_n - D|$ we mean the divisors $E$ of $g^r_n$ containing $D$, minus $D$.

$$|g^r_n - D| = \{E - D \mid E \in g^r_n, D \subseteq E\}.$$

If the degree of $D$ is $m$ then $(g^r_n - D) = g^{r-t}_{n-m}$ and we say

D imposes t (independent, linear) conditions on $g^r_n$. If $|D| = g^s_m$ then we would write

$$g^r_n = g^s_m + g^{r-t}_{n-m}$$

since any divisor in $g^s_m$ will impose t conditions on $g^r_n$. (Of course, we are assuming some divisor in $g^r_n$ contains D.)

As a final remark, when considering linear series one should always keep in mind the three possibilities: simple or composite; complete or incomplete; base points or no base points.

1.6 Line bundles. We give now a very brief discussion of line bundles. We refer to [13] and [14] for a more complete discussion.

Let $\{(U_\alpha, \varphi) \mid \alpha \in A\}$ be a set of charts where $X_p = \bigcup U_\alpha$ and $\varphi_\alpha : U_\alpha \to \Delta_\alpha$ where $\Delta_\alpha$ is a simply connected open set (disk) in $\mathbb{C}$. A (holomorphic) line bundle, $\eta$, is a collection of functions called transition functions $\{g_{\alpha\beta} \mid \alpha, \beta \in A, U_\alpha \cap U_\beta = \emptyset\}$ so that

(1) $g_{\alpha\beta} : U_\alpha \cap U_\beta \to \mathbb{C} - \{0\}$ is holomorphic

(2) $g_{\alpha\alpha} \equiv 1$

(3) $g_{\alpha\beta}\, g_{\beta\gamma}\, g_{\gamma\alpha} \equiv 1$ or $g_{\alpha\beta}\, g_{\beta\gamma} = g_{\alpha\gamma}$.

Example: The canonical line bundle. If for $\alpha$ and $\beta$ we have the following diagram

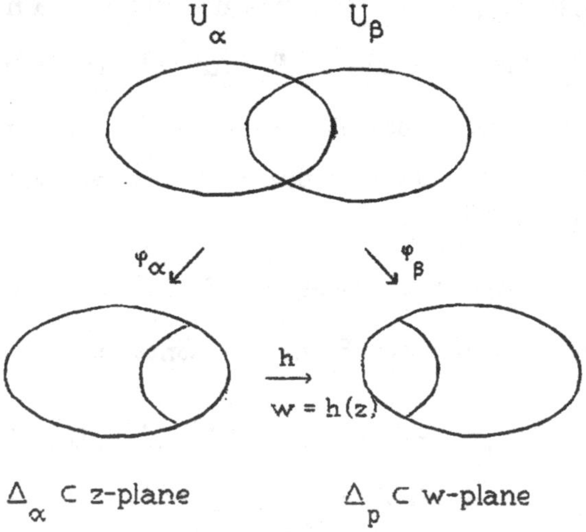

then $g_{\alpha\beta} = \frac{dh}{dz} \circ \varphi_\alpha$. The canonical line bundle will also be denoted by $K$.

A holomorphic (meromorphic) section of a line bundle $\eta$ is a set of holomorphic (meromorphic) functions $\{s_\alpha \mid \alpha \in A\}$ $s_\alpha : U_\alpha \to \mathbb{C}$ (or $\mathbb{P}^1$) so that

$$s_\alpha = g_{\alpha\beta} s_\beta \text{ in } U_\alpha \cap U_\beta .$$

The set of sections of a lie bundle form a vector space over $\mathbb{C}$. For example, the sections of the canonical bundle are differentials.

Theorem. On a compact Riemann surface the vector space of holomorphic sections of a holomorphic line bundle is finite dimensional.

Two line bundles $\eta = \{g_{\alpha\beta}\}$ and $\nu = \{h_{\alpha\beta}\}$ are said to be equivalent if there exists a set of non-zero holomorphic functions $\{t_\alpha \mid \alpha \in A\}$ and

$$t_\alpha g_{\alpha\beta} = h_{\alpha\beta} t_\beta .$$

Thus if $\{s_\alpha\}$ is a section of $\eta$ then $\{t_\alpha s_\alpha\}$ is a section of $\nu$.

Thus the sections of equivalent line bundles can appear to be quite different. What is invariant?

(a) If $\{s_\alpha\}$ is a section of a line bundle then the divisor of $\{s_\alpha\}$ makes sense since the transition functions are non-zero holomorphic functions. Let $(\{s_\alpha\})$ denote the divisor of the section $\{s_\alpha\}$. Then the set of divisors of holomorphic sections of a line bundle is independent of the equivalence class of the line bundle.

(b) If $\{s_\alpha\}$ and $\{s'_\alpha\}$ are two holomorphic sections of a line bundle $\eta$ then $\{s_\alpha/s'_\alpha\}$ defines a global meromorphic function. Therefore $(\{s_\alpha\}) \equiv (\{s'_\alpha\})$. So the divisors of holomorphic sections of a line bundle form a linear series which is easily seen to be complete.

(c) If $s_0, \cdots, s_r$ are $r+1$ linearly independent sections of $\eta$ then we can map $X \to \mathbb{P}^r$ by

$$x \to (s_0(x), \cdots, s_r(x)) .$$

(The curve in $\mathbb{P}^r$ is determined up to a projective transformation of $\mathbb{P}^r$ corresponding to different choices of bases in $\langle s_0, \cdots, s_r \rangle$, the span of those sections.) The curve is independent of the equivalence class of $\eta$.

(d) If $D$ is a divisor on $X$ we form a line bundle $[D]$ as follows. In

$U_\alpha$ let $f_\alpha$ be a meromorphic function whose divisor is $D \cap U_\alpha$. Let $g_{\alpha\beta} = f_\alpha / f_\beta$ in $U_\alpha \cap U_\beta$, a non-zero holomorphic function. (If $D \cap U_\alpha = 0$ let $f_\alpha = 1$.) Then clearly $\{f_\alpha\}$ is a meromorphic section of $\{g_{\alpha\beta}\}$ $\left(= [D]\right)$, whose divisor is $D$. Any holomorphic section of $[D]$ will have a positive divisor equivalent to $D$. Thus vector spaces of holomorphic sections of $[D]$ will parametrize linear series of divisors equivalent to $D$.

(e) There is a one-to-one correspondence between complete linear series of projective dimension greater than or equal to zero and holomorphic line bundles admitting holomorphic sections, given by $|D| \longleftrightarrow [D]$.

If $\eta = \{g_{\alpha\beta}\}$ and $\nu = \{h_{\alpha\beta}\}$ are line bundles then $\{g_{\alpha\beta} h_{\alpha\beta}\}$ and $\{g^{-1}_{\alpha\beta}\}$ are also line bundles, the former denoted $\eta\nu$ (or $\eta \otimes \nu$) called the (tensor) <u>product</u> of $\eta$ and $\nu$, and the latter denoted $\eta^{-1}$ called the <u>inverse</u> of $\eta$. If $K$ is the canonical line bundle then $K \otimes K \otimes \cdots \otimes K$ ($k$ times) or $K^k$, will be called the <u>k-canonical line bundle</u> (or written additively, $kK$) whose sections are called $k$-canonical differentials. With respect to a local parameter $z$ a $k$-canonical differential is written $f(z)(dz)^k$ and the transition functions for $kK$ are $\left(\frac{dh}{dz}\right)^k \circ \varphi_\alpha$, referring to the diagram earlier in this section.

## Chapter 2. Some exceptional points on Riemann surfaces.

2.1. On a Riemann surface, $X_p$, the points where a holomorphic differential has a zero of order greater than $p-1$ are called Weierstrass points. There are only a finite number of such points, at most $(p-1)\,p\,(p+1)$. If there are fewer then to each Weierstrass point can be associated a weight and the sum of all the weights is again $(p-1)\,p\,(p+1)$. The maximum weight at any point is $p(p-1)/2$, and if the maximum weight is attained, then the Riemann surface is hyperelliptic. In this case the Weierstrass point is a fixed point for the hyperelliptic involution, that is, a branch point for the unique $g^1_2$ on $X_p$. The purpose of this chapter is to extend to more general linear series these results of Hurwitz [15].

2.2. Let $g^r_n$ be a linear series without base points but not necessarily simple or complete. Let $\langle s_0, s_1, \cdots, s_r\rangle$ be a basis for the sections of the line bundle L determined by $g^r_n$ which generates the divisors in $g^r_n$. Fix $x_0$ in X. In a local parameter z at $x_0$, $z(x_0) = 0$, consider the Wronskian $W(s_0, \cdots, s_r) =$

$$\begin{vmatrix} s_0(z) & s_1(z) & \cdots & s_r(z) \\ s'_0(s) & s'_1(z) & \cdots & s'_r(z) \\ \cdot & \cdot & \cdots & \cdot \\ \cdot & \cdot & \cdots & \cdot \\ \cdot & \cdot & \cdots & \cdot \\ s^{(r)}_0(z) & s^{(r)}_1(z) & \cdots & s^{(r)}_r(z) \end{vmatrix}$$

where $s^{(k)}(z) = \frac{d^k s}{dz^k}(z)$, the $k^{th}$ derivative of s with respect to the local parameter z. If we choose a different basis $\langle t_0, \cdots, t_r\rangle$ the two bases are related by a constant matrix whose determinant is A and

$$W(s_0, \cdots, s_r) = A\,W(t_0, \cdots, t_r).$$

Thus the order of vanishing of $W(s_0, \cdots, s_r)$ at $x_0$ is independent of the basis chosen for $g^r_n$.

2.3. We now choose a basis for the sections <u>adapted to $x_0$</u> as follows. Using the local parameter z, $z(x_0) = 0$, we will find integers

$$0 = n_0 < n_1 < n_2 < \cdots < n_r$$

and a basis of sections $\langle t_0, \cdots, t_r \rangle$ so that $t_j$ has order of vanishing at $x_0$ equal to $n_j$. $\langle t_0, \cdots, t_r \rangle$ will be a basis adapted to $x_0$ and the order of vanishing of $W(t_0, \cdots, t_r)$ at $x_0$ will be

$\sum_{j=0}^{r} (n_j - j)$. This last sum will be seen to be independent of the various choices made.

Let $n_0 = 0$, and let $t_0$ be a section not vanishing at $x_0$ (since $g^r_n$ is without base points). Let $E_0 = (t_0)$ a divisor disjoint from $x_0$.

Let $S_1 = \{\text{sections } s \mid s(0) = 0\}$. Let $n_1$ be the smallest order of vanishing at $0$ of sections in $S_1$. $n_1 \geq 1$. Choose a section $t_1$ which vanishes precisely to order $n_1$ at $0$.

Then $\quad (t_1) = n_1 x_0 + E_1$, $(x_0, E_1) = 0$

and $\quad E_1 \in (g^r_n - n_1 x_0)$ $(= g^{r-1}_{n-n_1})$.

Let $S_2 = \{\text{sections } s \mid s^{(n_1)}(0) = 0\}$. Let $n_2$ be the smallest order of vanishing at $0$ for $s \in S_2$. Choose $t_2 \in S_2$ which vanishes to order $n_2$ at $z = 0$. $n_2 > n_1$.

Then $\quad (t_2) = n_2 x_0 + E_2$, $(x_0, E_2) = 0$

and $\quad E_2 \in (g^r_n - n_2 x_0)$ $(= g^{r-2}_{n-n_2})$.

Continue. We obtain sections $\langle t_0, \cdots, t_r \rangle$ and integers $0 = n_0 < n_1 < n_2 < \cdots < n_r$ where

(i) $\quad (t_j) = n_j x_0 + E_j$, $(x_0, E_j) = 0$

and $\quad E_j \in (g^r_n - n_j x_0)$ $(= g^{r-j}_{n-n_j})$.

(ii) in the local parameter $z$

$$t_j(z) = z^{n_j} g_j(z), \; g_j(0) \neq 0.$$

2.4. We now show the order of vanishng of $W(t_0, \cdots, t_r)$ at $x_0$.

$(z = 0)$, is $\sum_{j=0}^{r} (n_j - j)$. We follow Farkas-Kra [11, p. 78].

First we note that if $g$ is holomorphic in a neighborhood of $x_0$ then $W(gt_0, gt_1, \cdots, gt_r) = g^{r+1} W(t_0, t_1, \cdots, t_r)$. For a typical column of the matrix whose determinant is $W(gt_0, \cdots, gt_r)$ is

$$^t(gt_i\,,\, g't_i + gt'_i\,,\, g''t_i + 2g't'_i + gt''_i\,,\, \cdots)\,.$$

Using elementary row operations this reduces to

$$^t(gt_i\,,\, gt'_i\,,\, gt''_i\,,\, \cdots)$$

and the result follows. It now follows that the order of vanishing of $W(s_0, \cdots, s_r)$ at $x_0$ does not depend on the local parameter.

Now to prove the order of vanishing of $W(t_0, \cdots, t_r)$ is

$\sum (n_j - j)$ we proceed by induction on $r$. For $r = 0$ the result is immediate. Suppose it is true for integers less than $r$.

$$W\left(t_0, t_1, \cdots, t_r\right) = t_0^{r+1}\, W\left(1, \frac{t_1}{t_0}, \cdots, \frac{t_r}{t_0}\right)$$

$$= t_0^{r+1}\, W\left(\left(\frac{t_1}{t_0}\right)', \cdots, \left(\frac{t_r}{t_0}\right)'\right).$$

Now $\frac{t_j}{t_0} = z^{n_j} h_j(z)$ where $h_j(0) \neq 0$

$$\left(\frac{t_j}{t_0}\right)' = z^{n_j} h'_j + n_j z^{n_j - 1} h_j$$

$$= z^{n_1 - 1} \left(z^{n_j - n_1}\right)(z h'_j + n_j h_j)\,.$$

Therefore $W(t_0, \cdots, t_r) = (z^{n_1 - 1})^r\, W(f_1, \cdots, f_r)$ where

$f_j = z^{n_j - n_1} g_j$ and $g_j(0) \neq 0$. By induction $W(f_1, \cdots, f_r)$ vanishes to order

$$\sum_{j=1}^{r} (n_j - n_1) - (j - 1) .$$

Thus $W(t_0,\cdots,t_r)$ vanishes to order

$$r(n_1 - 1) + \sum_{j=1}^{r} (n_j - n_1 - j + 1)$$

$$= \sum_{j=0}^{r} n_j - j \quad (n_0 = 0) .$$

2.5. We now show that $W(s_0,\cdots,s_r)$ is a section of the line bundle $L^{r+1}K^{r(r+1)/2}$ where $L$ is the line bundle corresponding to the completion of $g^r_n$ and $K$ is the canonical line bundle.

In $U_z \cap V_w$, the intersection of two parametric neighborhoods of $x_0$ with local parameters $z$ and $w$ let $g$ be the transition function for $L$. If $s_i$ is a section, let $f_i(z)$ and $h_i(w)$ be the representatives of $s_i$ in $U_z$ and $V_w$ respectively where $w = w(z)$ is biholomorphic. Then

$$f_i(z) = g(z)h_i\big(w(z)\big)$$

$$W(f_0,\cdots,f_r) = W(gh_0,\cdots,gh_r) = g^{r+1}W\Big(h_0\big(w(z)\big),\cdots,h_r\big(w(z)\big)\Big) .$$

Now

$$\frac{d}{dz}h_i\big(w(z)\big) = \frac{dh_i}{dw}\cdot w'(z)$$

$$\frac{d^2}{dz^2}h_i\big(w(z)\big) = \frac{d^2h_i}{dw^2}\cdot (w')^2 + \frac{dh_i}{dw}\cdot w''$$

$$\frac{d^3}{dz^3}h_i\big(w(z)\big) = \frac{d^3h_i}{dw^3}\cdot (w')^3 + 3\frac{d^2h_i}{dw^2}\cdot w'w'' + \frac{dh_i}{dw}\cdot w''' .$$

Again by elementary row operations we see that the $i^{th}$ column in the determinant becomes

$${}^t\Big(h_i , \frac{dh_i}{dw}\cdot w' , \frac{d^2h_i}{dw^2}\cdot (w')^2 , \frac{d^3h_i}{dw^3} , (w')^3 , \cdots\Big) .$$

Thus we see that

$$W(f_0,\cdots,f_r) = g^{r+1}(w')^{r(r+1)/2}\,W(h_0,\cdots,h_r) .$$

Since $g^{r+1}(w')^{r(r+1)/2}$ is the transition function for $L^{r+1}K^{r(r+1)/2}$ the proof is complete.

It follows that the degree of $W(s_0,\cdots,s_r)$ is

$$(r+1)n + \big(r(r+1)/2\big)(2p-2)$$
$$= (r+1)\big(n + r(p-1)\big) \qquad [9, \text{p. } 132].$$

2.6. We now make some definitions and summarize the results of the last several sections in a theorem.

Let $g^r_n$ be a linear series (without base points) which is a subseries of the complete linear series corresponding to a line bundle $L$. A point $x \in X_p$ where a divisor of $g^r_n$ has order greater than $r$ will be called a generalized Weierstrass point for $g^r_n$ The Weierstrass weight of $g^r_n$ at $x$, denoted $w(g^r_n, x)$, is computed as in Section 2.4.

Theorem 2.6. The total sum $\sum_{x \in X_p} w(g^r_n, x)$ is equal to $(r+1)\big(n + r(p-1)\big)$. The divisor of generalized Weierstrass points for $g^r_n$ (counting multiplicities) is a divisor corresponding to the line bundle $L^{r+1}K^{r(r+1)/2}$.

If $g^r_n$ is complete and non-special this total is $(r+1)^2 p$. In particular, if $g^r_n$ is the k-canonical series, which we denote by $kK$ $(k \geq 2)$, then this total is $(2k-1)^2(p-1)^2 p$. A generalized Weierstrass point for the k-canonical series will be called a higher order Weierstrass point. It is known that the higher order Weierstrass points are dense in $X_p$ [22].

If $g^r_n = |G|$, a complete linear series, we will often write $w(G,x)$ for $w(g^r_n, x)$.

For a linear series $g^r_n$ and $x \in X$, the integers $n_0, n_1, \cdots, n_r$ will be called the non-gaps for $g^r_n$ at $x$. The complimentary integers in $[0,n]$ will be called the gaps for $g^r_n$ at $x$. (For the canonical series note that this terminology differs from the usual terminology.) Then

$$w(g^r_n, x) = \sum_{j=0}^{r} (n_j - j) = \sum_{j=1}^{n-r} (r + j - g_j)$$

where $g_1, \cdots, g_{n-r}$ are the gaps for $g^r_n$ at $x$. If we let $\hat{n}_i$ denote a non-gap greater than $r$ (an extraordinary non-gap) and let $\hat{g}_i$ denote a gap less than $r+1$ (an extraordinary gap) then

$$w(g^r_n, x) = \sum \hat{n}_i - \hat{g}_i$$

and the number of extraordinary gaps equals the number of extraordinary non-gaps.

2.7 Examples. The generalized Weierstrass points for $g^1_n$ are the branch points for the cover $X_p \to \mathbb{P}^1$ given by $g^1_n$. The number is $2(n - p + 1)$ which agrees with the Riemann-Hurwitz formula.

For a simple $g^2_n$ the inflection points (flexes) of the corresponding plane model, $C_n$, are among the generalized Weierstrass points. However, one sees that cusps also contribute to the total, in particular, a simple cusp contributes 2 to the total. If the plane curve is non-singular then the total weight is $3\big(n + 2(p-1)\big) = 3n(n-2)$ since $p = (n-1)(n-2)/2$, and this counts the number of inflection points.

For $p = 3$ and $X_3$ non-hyperelliptic, the $C_4$ corresponding to the canonical $g^2_4$ has 24 flexes (counting multiplicities).

2.8 Theorem. Let $g^r_n$ be a complete linear series without base points. The maximum weight at a point of $X_p$ is $(r = n - p + i)$

$$\frac{(n-r)(n-r+1)}{2} = \frac{(p-i)(p-i+1)}{2}$$

The maximum is attained only on hyperelliptic Riemann surfaces and only at hyperelliptic Weierstrass points.

Remarks. For $g^r_n = g^{p-1}_{2p-2}$ this gives Hurwitz's result. The method of proof of the theorem is that of Segre [24], [9, p. 292].

Proof. For $x_0 \in X_p$ we find a basis of sections for a line bundle corresponding to $g^r_n$ adapted to $x_0$ as in Section 2.3. We obtain

$$0 = n_0 < n_1 \cdots < n_r$$

so that we have divisors in $g^r_n$:

$$n_j x_0 + E_j$$

where $(x_0, E_j) = 0$, $|E_j| = g^{r-j}_{n-n_j}$, $j = 0, 1, \cdots, r$.

Let $s$ be the smallest integer so that $n_s > s$. Then $s \geq 1$ and for $j \geq s$, $g^{r-j}_{n-n_j}$ is special since

$$(n - n_j) - (r - j) = n - r - (n_j - j) < n - r = p - i \leq p.$$

By Clifford's theorem for $j \geq s$

$$n - n_j \geq 2r - 2j$$

or

$$n - 2r + j \geq n_j - j.$$

Now $n_s \geq s + 1$ so $n - 2r + s \geq n_s - s \geq 1$ and $s \geq 2r - n + 1$.

Computing: $w(g^r_n, x_0) = \sum_{j=s}^{r} (n_j - j)$

$$\leq \sum_{j=2r-n+1}^{r} (n - 2r + j)$$

$$= \sum_{k=1}^{n-r} k = \frac{(n-r)(n-r+1)}{2}.$$

If we have equality in the last formula we have equality in all the preceeding inequalities, and so $X_p$ hyperelliptic by the strong form of Clifford's theorem. In particular if we let $j = r - 1$

then $\quad n - n_{r-1} = 2r - 2r + 2 = 2$

and $\quad n - n_r = 0$.

Consequently $|E_{r-1}| = g^1_2$ and $g^r_n = |n x_0|$.

It follows that $2x_0 \in g^1_2$ and so $x_0$ is a hyperelliptic Weierstrass point.

q.e.d.

Chapter 3. The inequality of Castelnuovo-Severi.

3.1. As motivation for this inequality we consider for hyperelliptic Riemann surfaces some properties which are part of the folklore of the subject. These properties will be generalized by the Castelnuovo-Severi inequality and also by some later results on automorphisms.

3.2 Lemma. Suppose $X_p$ is hyperelliptic and $f \in M(X_p)$. Let $\pi : X_p \to \mathbb{P}^1$ be a 2-sheeted cover. (Then $\pi$ is a meromorphic function of order 2.) Suppose $o(f) \leq p$. Then there exists a rational function $R(X) \in \mathbb{C}(X)$ and $f = R \circ \pi$ and so $o(f)$ is even.

Proof. We may assume that no poles of $f$ lie on the branch points of $\pi$. For if $T$ is a fractional linear transformation then $T \circ f$ will have the same properties as $f$ except that by choosing $T$ properly the poles of $T \circ f$ can be moved away from any finite set on $X_p$. If the conclusion of the lemma holds for $T \circ f$ it will hold for $f$.

Let $B$ be the image of the $2p + 2$ branch points of $\pi$. For $z \in \mathbb{P}^1 - B$, let $\pi^{-1}(z) = \{x_1, x_2\}$ and let $g(z) = \big(f(x_1) - f(x_2)\big)^2$. Then $g$ is meromorphic on $\mathbb{P}^1 - B$ and bounded near $B$ and so can be extended to be meromorphic on all of $\mathbb{P}^1$. $g$, so extended, has at most $2\,o(f)$ poles, and at least $2p + 2$ zeros at the points of $B$. If $g(z)$ is not identically zero then

$$2\,o(f) \geq \#\text{ of poles of } g = \#\text{ of zeros of } g \geq 2p + 2 .$$

But $o(f) \leq p$, so $g$ is identically zero. That is, $f$ identifies the fibers of $\pi$, and so $f$ is the lift of a function $R$ from $\mathbb{P}^1$. $f = R \circ \pi$ where $R$ is rational.

q.e.d.

3.3 Lemma. For $p \geq 2$, the $g^1_2$ on a hyperelliptic $X_p$ is unique.

Proof. A $g^1_2$ is the set of fibers of a meromorphic function of order 2. If $f$ has order 2 then $f = R \circ \pi$ where $\pi$ is the original function of order 2. Thus $R$ must have order one in $M(\mathbb{P}')$; that is, $R$ is a fractional linear transformation. Consequently $\pi$ and $f$ have the same set of fibers as maps from $X_p \to \mathbb{P}^1$.

q.e.d.

3.4. Here is another result that asks to be generalized but which will not be considered further since the methods considered in these notes do not seem entirely appropriate.

Lemma 3.4. Suppose $X_p$ is hyperelliptic and $\pi : X_p \to Y_q$ is a map of compact Riemann surfaces. If $q \geq 2$ then $Y_q$ is hyperelliptic.

Proof. Let $f$ be a function of order 2 on $X_p$. If the map $X_p \to Y_q$ is n-sheeted, for almost all $y$ in $Y_q$ let $\pi^{-1}(y) = \{x_1, \cdots, x_n\}$. Define

$$g(y) = f(x_1) f(x_2) \cdots f(x_n)$$

which can be extended to be a well defined meromorphic function on $Y$. If zeros and poles of $f$ happen to lie in the same fiber of the map $\pi$, replace $f$ by $f - \lambda_0$ for some $\lambda_0 \in \mathbb{C}$. Then $g(y)$ will be a function of order 2 on $Y_q$.

q.e.d.

3.5 Statement of the inequality of Catelnuovo-Serveri.

Theorem 3.5. [8].

(i) Suppose we have three compact Riemann surfaces related in the following manner

$$\begin{array}{ccccc} & & X_p & & \\ & \swarrow^{\pi} & & {}^{\rho}\searrow & \\ X_q & & & & X_r \end{array}$$

where the holomorphic maps $\pi$ and $\rho$ have $m$ and $n$ sheeets respectively.

(ii) Suppose further that the diagram admits no proper factorization; that is, there does not exist an $X_{p'}$ with $p' < p$ so that we have the following diagram

$$\begin{array}{ccccc} & & X_p & & \\ & & \downarrow & & \\ & & X_{p'} & & \\ & \swarrow & & \searrow & \\ X_q & & & & X_r \end{array}.$$

(iii) Then $p \leq mq + nr + (m-1)(n-1)$.

(iv) On $X_q$ we have divisors of degree $n$

$$\{\pi \circ \rho^{-1}(y) \,|\, y \in X_r\}$$

called an algebraic series of degree $n$ and index $m$, again

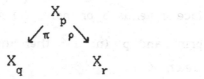

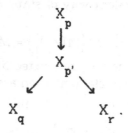

denoted $\gamma^1_n$. (If $m = 1$, $\gamma^1_n$ is an involution of genus $r$.)
In general $m$ divisors of $\gamma^1_n$ contain the same point $x_0 \in X_q$.

(v) If we have equality in (iii) above, then all the divisors in $\gamma^1_n$ are linearly equivalent (but do not necessarily form a linear series.)

3.6 Applications. Many applications are based on the fact that if the inequality in (iii) does not hold then (ii) is false; that is, the diagram does then admit a proper factorization.

(i) If $q = r = 0$, $m = 2$ and $n \leq p$, then the results of Lemma 3.2 follow.

(ii) A Riemann surface of genus greater than 3 admitting a meromorphic function of order 3 is called trigonal. With $q = r = 0$, $m = 3$ and $2n < p$, we see that any meromorphic function $f$ on a trigonal Riemann surface of order less than $p/2$ is a rational function of a function of order 3 and also $3 \mid o(f)$. On a trigonal Riemann surface of genus 5 or more, the $g^1_3$ is unique.

(iii) If $n$ is prime and $p > (n-1)^2$ then any $g^1_n$ without base points is unique on $X_p$.

(iv) Suppose $p > 9$ and $X_p$ admits two complete $g^1_4$'s without base points. Then with $q = r = 0$ and $m = n = 4$ we see that the diagram admits a proper factorization and $X_{p'}$ admits two complete $g^1_2$'s. Consequently $p' = 1$. Thus $X_p$ is elliptic-hyperelliptic; that is, a 2-sheeted covering of a Riemann surface of genus one.

3.7 Some remarks. An equivalent way of stating hypothesis (ii) of Theorem 3.5 is the following.

$$\left[\pi^*\left(M(X_q)\right), \rho^*\left(M(X_r)\right)\right] = M(X_p)$$

where the bracket indicates the field generated by the two fields inside.

The theorem of the primitive element then assures us that

$$M(X_p) = \rho^*\left(M(X_r)\right)[g]$$

where $g$ is some function in $\pi^*\left(M(X_q)\right)$. It follows that for almost all $y \in X_r$, $\rho^{-1}(y)$ is a set of $n$ distinct points on which $g$ takes $n$ distinct values. From this it follows that almost all divisors in $\gamma^1_n$ are composed of $n$ distinct points.

3.8 Proof of Theorem 3.5.

Let $g^{n-1}_{n-1+q}$ be a complete, non-special linear series on $X_q$, without base points. If $|D| = g^{n-1}_{n-1+q}$ let $f_1, \cdots, f_n$ be a basis of $L(D)$. $g^{n-1}_{n-1+q}$ lifts to $X_p$ to become a $g^{n-1}_{m(n-1+q)}$ ( no longer necessarily complete). Let $\hat{f}_i = f_i \circ \pi$. If $y \in X_r$ let $\rho^{-1}(y) = \{x_1, \cdots, x_n\}$ and define $h(y)$ as follows.

$$h(y) = \begin{vmatrix} \hat{f}_1(x_1) \cdots\cdot \hat{f}_1(x_n) \\ \cdot \\ \cdot \\ \cdot \\ \cdot \\ \hat{f}_n(x_1) \cdots\cdot \hat{f}_n(x_n) \end{vmatrix}^2 = \begin{vmatrix} f_1(\pi x_1) \cdots\cdot f_1(\pi x_n) \\ \cdot \\ \cdot \\ \cdot \\ \cdot \\ f_n(\pi x_1) \cdots\cdot f_n(\pi x_n) \end{vmatrix}^2$$

We prove the inequality in 3 steps:

(i) We show $g^{n-1}_{n-1+q}$ can be chosen so $h(y)$ is not identically zero.

(ii) We then show that the number of poles of $h(y)$ is less than or equal to $2m(n-1+q)$.

(iii) We show the numbers of zeros of $h(y)$ is greater than or equal to $\operatorname{ram} \rho = 2p - 2 - n(2r - 2)$.

Since the number of poles of $h$ equals the number of zeros we have

$$2m(n - 1 + q) \geq 2p - 2 - n(2r - 2)$$

which is the inequality of Castelnuovo-Severi.

(i) Suppose that $h(y)$ is identically zero $\big(h(y) \equiv 0\big)$. Then every divisor in $g^1_n$ must lie in a divisor of $g^{n-1}_{n-1+q}$. For if $h(y) = 0$, $\rho^{-1}(y) = \{x_1, \cdots, x_n\}$, then there exist $\lambda_1, \cdots, \lambda_n \in \mathbb{C}$, not all zero, and $\sum_{j=1}^{n} \lambda_j f_j(\pi x_i) = 0$ for all $i$. Thus $\{\pi x_1, \cdots, \pi x_n\}$ is part of the divisor

$$D_\lambda = \left\{ z \in X_q \,\middle|\, \sum \lambda_j f_j(z) = 0 \right\}.$$

Now fix $y_0 \in X_r$ so that $\pi \circ \rho^{-1}(y_0)$ $\left(= \{\pi x_1, \cdots, \pi x_n\}\right)$ is a set of $n$ distinct points, and no $x_i$ is a branch point of $\rho$. Let $D_q$ be a non-special integral divisor on $X_q$ of degree $q$, and

$\pi x_n \notin D_q$ ; $|D_q| = g^0_q$ . Let $|D_q + \pi x_1 + \cdots + \pi x_{n-1}| = g^r_{n-1+q}$ . Since this linear series is non-special $r = n - 1$. Moreover, $g^r_{n-1+q} - \pi x_1 - \cdots - \pi x_{n-1} = D_q$. Thus no divisor of $g^r_{n-1+q}$ contains $\pi x_1 + \cdots + \pi x_{n-1} + \pi x_n$ .

(ii) and (iii) We will combine these proofs in a lemma.

<u>Lemma 3.8.1</u>. Let $\rho : X_p \to X_r$ be an $n$-sheeted covering of compact Riemann surfaces $(n \geq 2)$. Let $D$ be a divisor on $X_p$ so that $|D| = g^R_N$ where $R \geq n - 1$. Let $f_1, \ldots, f_n$ be $n$ independent functions in $L(D)$. For $y \in X_r$, let $\rho^{-1}(y) = \{x_1, \cdots, x_n\}$. Define $h(y)$ by

$$h(y) = \begin{vmatrix} f_1(x_1) & f_1(x_n) \\ \cdot & \cdot \\ \cdot & \cdot \\ \cdot & \cdot \\ f_n(x_1) & f_n(x_n) \end{vmatrix}^2$$

If $h(y)$ is not identically zero then $N \geq p - 1 - n(r - 1)$.

<u>Proof</u>. Since $R \geq 1$ we can assume that $D$ is a divisor of $N$ distinct points which are disjoint from the set of branch points of $\rho$. Then if $k$ points of $D$ are in $\rho^{-1}(y)$ one sees that $h$ has a pole of order at most $2k$ at $y$. Consequently $h$ has at most $2N$ poles.

Now we show that if $x_0 \in X_p$ and $\mathrm{ram}_{x_0} \rho = s$ then $x_0$ contributes at least $s - 1$ to the zeros of $h(y)$.

Choose a basis of $\langle f_1, \cdots, f_n \rangle$ adapted to $x_0$ (Section 2.3). Call this new basis $\langle h_1, \cdots, h_n \rangle$. In a parametric neighborhood of $x_0$, $U$, with local parameter $t$ $\left(t(x_0) = 0\right)$ the map $\rho$ is equivalent to $t \to t^s = z$, where $z$ is a local parameter at $\rho(x_0) \in X_r$ $\left(z(\rho(x_0)) = 0\right)$. Using these local parmeters we have

$$\rho^{-1}(z) \cap U = \{t, t\tau, \cdots, t\tau^{s-1}\}$$

where $\tau = e^{2\pi i/s}$ and $t^s = z$. Then

$$\begin{aligned} h_1(t) &= t^{\ell_1} k_1(t) \\ h_2(t) &= t^{\ell_2} k_2(t) \\ &\vdots \\ h_n(t) &= t^{\ell_n} k_n(t) \end{aligned}$$

where $0 \le \ell_1 < \ell_2 < \cdots < \ell_n$, $j \le \ell_j$ and $k_j(0) \ne 0$. Using these parameters we have

$$h(z) = A^2 \begin{vmatrix} h_1(t) & h_1(\tau t) & h_1(\tau^{s-1}t) \\ h_2(t) & h_2(\tau t) & h_2(\tau^{s-1}\tau) \\ \vdots & & \\ h_s(t) & h_s(\tau t) & h_s(\tau^{s-1}t) \\ h_{s+1}(t) & h_{s+1}(\tau t) & h_{s+1}(\tau^{s-1}t) \\ \vdots & & \\ h_n(t) & h_n(\tau t) & h_n(\tau^{s-1}t) \end{vmatrix}^2 \quad (A \ne 0)$$

Each $s \times s$ determinant from the first $s$ rows has a zero (as a function of $t$) of order at least $0 + 1 + 2 + \cdots + (s-1) = s(s-1)/2$. Thus as a function of $t$ the square of the determinant is divisible by $t^{s(s-1)}$. But $t^s = z$, so $h$ has a zero of order at least $s - 1$ coming from the first $s$ rows of the determinant.

Thus $h(y)$ has at least $\operatorname{ram}\rho$ zeros and $\operatorname{ram}\rho = 2p - 2 - n(2r - 2)$. Since $h(y)$ has an equal number of zeros and poles the result follows.

q.e.d.

This completes the proof of the Castelnuovo-Severi inequality.

Now suppose we have equality. Then every inequality in the proof becomes an equality. In this case the zeros of $h(y)$ arise only from the ramification points of $\rho$. Any $g^{n-1}_{n-1+q}$ for which the corresponding $h(y)$ is not identically zero has no divisor containing a divisor of $\gamma^1_n$ for that would add another zero to $h(y)$. If $g^{n-1}_{n-1+q}$ is chosen so that one divisor of $\gamma^1_n$ is in one of its divisors then <u>every</u> divisor of $\gamma^1_n$ lies in some divisor of $g^{n-1}_{n-1+q}$ since $h(y) \equiv 0$.

Now pick $D_n \in \gamma^1_n$ so that $D_n$ has distinct points. Choose $G$ to be a non-special divisor of distinct points of degree $q$.

Then $|G + D_n| = g^n_{n+q}$ and we can assume it is without base points. Since $|G| = g^0_q$, the $n$ points of $D_n$ impose independent conditions on $g^n_{n+q}$; that is, if

$$D_n = x_1 + x_2 + \cdots + x_n \quad \text{then}$$

$$g^n_{n+q} - x_{i_1} - x_{i_2} - \cdots - x_{i_t} = g^{n-t}_{n+q-t}, \; 1 \le t \le n .$$

Now $g^n_{n+q}$ gives a map $\Theta : X_p \to \mathbb{P}^n$. The statement that $D_n$ imposes independent conditions on $g^n_{n+q}$ translates into the statement: $\Theta(x_1),\cdots,\Theta(x_n)$ span a linear space of dimension $n-1$ in $\mathbb{P}^n$. This is an open condition; that is, for $E$ close to $D$ $(\deg E = n, E \ge 0)$ $E$ will also impose independent conditions on $g^n_{n+q}$.

<u>Lemma 3.8.2</u>. (Castelnuovo.) If $E_n \in \gamma^1_n$, then $E_n$ imposes independent conditions on $g^n_{n+q}$ if and only if $E_n \equiv D_n$.

<u>Proof.</u> Let $G = P_1 + \cdots + P_q$, $q$ distinct points. Consider $g^n_{n+q} - P_1 = |D_n + P_2 + \cdots + P_q| = g^{n-1}_{n+q-1}$, which contains $D_n$. Consequently $g^{n-1}_{n+q-1}$ contains all divisors of $\gamma^1_n$. Suppose

$E_n \in \gamma^1_n$ imposes independent conditions on $g^n_{n+q}$.

$$g^{n-1}_{n+q-1} = |E_n + H_{q-1}|$$

Thus

$$g^n_{n+q} = |E_n + H_{q-1} + P_1|$$

$$g^n_{n+q} - E_n = |H_{q-1} + P_1| = g^0_q$$

since $E_n$ imposes independent conditions on $g^n_{n+q}$. Therefore $P_1 \in g^n_{n+q} - E_n$. But there was nothing unique about $P_1$. Thus $P_j \in g^n_{n+q} - E_n$ for all $j$.

$$g^n_{n+q} - E_n = P_1 + \cdots + P_q = G$$

$$g^n_{n+q} = |D_n + G| = |E_n + G|$$

or $D_n \equiv E_n$.

Conversely, if $D_n \equiv E_n$ then $g^n_{n+q} - E_n = |G| = g^0_q$, so $E_n$ imposes independent conditions on $g^n_{n+q}$.

q.e.d.

We complete the proof that equality in the Castelnuovo-Severi inequality implies all divisors in $\gamma^1_n$ are linearly equivalent by a connectivity argument.

Let $X^{(n)}_q$ = {all integral divisors of degree n on $X_q$} and let $\tilde{\gamma}^1_n$ = {divisors in $\gamma^1_n$ consisting of n distinct points}. Then $\tilde{\gamma}^1_n$ is a connected subset of $X^{(n)}_q$ when the latter is given the natural topology. Fix $D_n \in \tilde{\gamma}^1_n$. Let $S = \left\{E_n \in \tilde{\gamma}^1_n \,\middle|\, E_n \equiv D_n\right\}$. Then S is easily seen to be closed in $\tilde{\gamma}^1_n$. But the lemma of Castelnuovo shows that S is open, that is, $S = \tilde{\gamma}^1_n$. Since $\tilde{\gamma}^1_n = \gamma^1_n$ - {finite set of divisors} the proof is complete.

<u>3.9 Castelnuovo's Riemann-Roch theorem.</u>

<u>Theorem 3.9</u>. [9, p. 324], [6]. Suppose $\gamma^1_n$ is an algebraic series of

index one and genus $q$ on $X_p$; that is, $\gamma^1_n$ is the set of fibers of an $n$-sheeted map $X_p \to X_q$. Let $g^R_N$ be a complete linear series on $X_p$ where $R \geq n - 1$. Suppose

$$N - R < p - nq.$$

Then any divisor in $\gamma^1_n$ will not impose independent conditions on $g^R_N$; that is, if $D \in \gamma^1_n$ then

$$|g^R_N - D| = g^{R-n+1+\varepsilon}_{N-n}$$

where $\varepsilon \geq 0$.

<u>Remarks</u>. Notice that $g^R_N$ may be composite and that we may assume $g^R_N$ is without base points without loss of generality. Clearly $g^R_N$ must be special. For $q = 0$ the theorem is a consequence of the Riemann-Roch theorem.

<u>Proof</u>. For the proof we will assume $g^R_N$ is without base points. Also for technical convenience we will prove the theorem only for divisors $D \in \gamma^1_n$ made up of $n$ distinct points.

Let $D_n = x_1 + \cdots + x_n$ be a divisor in $\gamma^1_n$ made up of $n$ distinct points. Suppose $D_n$ imposes $n$ independent conditions on $g^R_N$. We will arrive at a contradiction.

Let $D_{n-1} = x_1 + \cdots + x_{n-1}$. If $R \geq n$ we can find a divisor $Y$ of $R - n + 1$ distinct points so that $D_{n-1} + Y$ imposes $R$ conditions on $g^R_N$, and the unique divisor in $g^R_N$ containing $Y + D_{n-1}$ does <u>not</u> contain $x_n$.

Now $|g^R_N - Y| = g^{n-1}_{N-R+n-1}$. Since

$$N - R + n - 1 < p - 1 - n(q - 1)$$

every divisor in $\gamma^1_n$ lies in some divisor of $g^{n-1}_{N-R+n-1}$ (Lemma 3.8.1). Let $D_{n-1} + E$ be a divisor in $g^{n-1}_{N-R+n-1}$ containing $D_n$. Then $x_n \in E$. But $Y + D_{n-1} + E$ is the unique divisor in $g^R_N$ containing $Y + D_{n-1}$, and so $x_n \notin E$. This contradiction proves the theorem for divisors in $\gamma^1_n$ of $n$ distinct points.

q.e.d.

3.10 An application of Castelnuovo's Riemann-Roch Theorem.

Theorem 3.10. Let $C_N$ be a plane model for a Riemann surface $X_p$ where

$$p = (N-1)(N-2)/2 - \delta .$$

Suppose we have integers $m \geq 2$ and $q \geq 0$ so that

$$\delta + mq < (N-m)(N-m-1)/2 .$$

Then $X_p$ does not admit a $\gamma^1_m$ of index one and genus $q$.

Proof. If $g^2_N$ is the linear series cut out on $C_N$ by the lines in $\mathbb{P}^2$, then curves of degree $m-1$ cut out a $g^\rho_\nu$,

$$|(m-1)\, g^2_N| = g^\rho_\nu$$

where $\nu = (m-1)N$ and $\rho = (m-1)(m+2)/2 + \varepsilon$, $\varepsilon \geq 0$.

Suppose $X_p$ admits such a $\gamma^1_m$.

$$\begin{aligned}\text{Now, } \nu - \rho &= (m-1)N - (m-1)(m+2)/2 - \varepsilon \\ &\leq (m-1)N - (m-1)(m+2)/2 .\end{aligned}$$

Since $p = (N-1)(N-2)/2 - \delta$, elementary algebra shows that

$$\nu - \rho < p - mq .$$

Thus a divisor of $m$ distinct points in $\gamma^1_m$, $x_1 + \cdots + x_m$, imposes at most $m-1$ conditions on $g^\rho_\nu$. Suppose that any divisor in $g^\rho_\nu$ containing $x_1 + \cdots + x_{m-1}$ must contain $x_m$. We can find lines in $\mathbb{P}^2$, $\ell_i = 0$ so that $\ell_i = 0$ contains $x_i$ but not $x_m$, $i = 1,2,\cdots,m-1$. $\ell_1 \ell_2 \cdots \ell_{m-1} = 0$ cuts out a divisor in $g^\rho_\nu$ containing $x_1 + \cdots + x_{m-1}$ but not containing $x_m$. This contradiction shows that $\gamma^1_m$ does not exist.

q.e.d.

This generalizes the fact that a non-singular plane curve of degree 4 or more is not hyperelliptic. ($\delta = 0$, $m = 2$, $q = 0$.)

## Chapter 4. Smooth and branched coverings of Riemann surfaces.

4.1. In the following discussion many standard facts will be stated without proof. (Reference: Ahlfors and Sario [4].)

Definition. A continuous map $f : Y \to X$ between two surfaces will be called a (smooth) covering if $f$ is a local homeomorphism and for every $x \in X$, there is a connected open neighborhood $\Delta$ of $x$ so that $f^{-1}(\Delta) = \bigcup_{i=1}^{n \text{ or } \infty} \Delta_i$ where the sets $\{\Delta_i\}$ are the components of $f^{-1}(\Delta)$ and $f : \Delta_i \to \Delta$ is a homeomorphism.

Definition: Suppose $f : Y \to X$ is a local homeomorphism between surfaces. Let $\gamma : [0,1] \to X$ be a continuous map (path), and suppose $\gamma(0) = x_0$. Suppose $\tilde{\gamma} : [0,1] \to Y$ is a path so that $\tilde{\gamma}(0) = y_0$, $f(y_0) = x_0$, and $f \circ \tilde{\gamma} = \gamma$. Then $\tilde{\gamma}$ will be called a lift of $\gamma$ starting at $y_0$.

Lemma. If $\tilde{\gamma}$ and $\tilde{\gamma}'$ are lifts of $\gamma$ starting at $y_0$ then $\tilde{\gamma} = \tilde{\gamma}'$ (uniqueness of lifts).

Lemma. If $f : Y \to X$ is a local homeomorphism between surfaces, then the following are equivalent:

(a) $f$ is a covering;

(b) If $\gamma : [0,1] \to X$ is a path and $f(y_0) = \gamma(0)$, then there exist a lift of $\gamma$ starting at $y_0$.

Monodromy Theorem. Let $f : Y \to X$ be a covering. Let $\gamma_0, \gamma_1$ be paths in $X$ so that $\gamma_0(0) = \gamma_1(0)$, $\gamma_0(1) = \gamma_1(1)$, and $\gamma_0$ is homotopic to $\gamma_1$. Let $\tilde{\gamma}_0$ and $\tilde{\gamma}_1$ be lifts of $\gamma_0$ and $\gamma_1$ starting at the same point above $\gamma_0(0)$. Then $\tilde{\gamma}_0$ is homotopic to $\tilde{\gamma}_1$, and in particular $\tilde{\gamma}_0(1) = \tilde{\gamma}_1(1)$.

Theorem. Let $f : Y \to X$ be a covering. Then $\operatorname{card}\{f^{-1}(x)\}$ is a constant function on $X$, possibly infinite. This integer (or $\infty$) is called the number of sheets of the covering.

### 4.2 Classification of smooth coverings.

Let $X$ be a surface, $x_0 \in X$. A triple $(Y, f, y_0)$ will be called a covering of $(X, x_0)$ if $f : Y \to X$ is a covering of surfaces, $y_0 \in Y$, and $f(y_0) = x_0$. If $(Y, f, y_0)$ and $(Z, g, z_0)$ are coverings of $(X, x_0)$, they will be said to be equivalent if there exists a homeomorphism $h : Y \to Z$ so that $h(y_0) = z_0$ and $g \circ h = f$.

Suppose $(Y,f,y_0)$ is a covering of $(X,x_0)$. Let $[\gamma]$ be an equivalence class of paths in $\pi_1(X,x_0)$. Let $D \subset \pi_1(X,x_0)$ be the paths $[\gamma]$ so that the lift of $\gamma$ starting at $y_0$ is a closed path. By the monodromy theorem $D$ is a well defined subgroup of $\pi_1(X,x_0)$. In fact if we define $f_* = \pi_1(Y,y_0) \to \pi_1(X,x_0)$ by $f_*[\gamma] = [f \circ \gamma]$ then $D = f_*(\pi_1(Y,y_0))$.

<u>Existence Theorem</u>. If $D$ is a subgroup of $\pi_1(X,x_0)$, there exists a covering $(Y,f,y_0)$ of $(X,x_0)$ so that $D = f_*(\pi_1(Y,y_0))$.

<u>Theorem</u>. Suppose $D \subset D' \subset \pi_1(X,x_0)$ and let $(Y,f,y_0)$ and $(Y',f',y'_0)$ be coverings as in the existence theorem. Then there exist a covering $g : Y \to Y'$ so that $(Y,g,y_0)$ is a covering of $(Y',y'_0)$ and $f = f' \circ g$.

<u>Corollary</u>. $(Y,f,y_0)$ and $(Y',f',y'_0)$ are equivalent coverings of $(X,x_0)$ if and only if $D = D'$.

Suppose for each subgroup $D \subset \pi_1(X,x_0)$ we pick one covering $(Y_D,f_D,y_D)$ as in the existence theorem. We define a partial ordering of these covers as follows: $(Y_D,f_D,y_D) > (Y_{D'},f_{D'},y_{D'})$ if there exist a covering $g : Y_D \to Y_{D'}$ so that $(Y_D,g,y_D)$ is a covering of $(Y_{D'},y_{D'})$ and $f_D = f_{D'} \circ g$. Then $(Y_D,f_D,y_D) > (Y_{D'},f_{D'},y_{D'})$ if and only if $D \subseteq D'$. Thus the map $D \to (Y_D,f_D,y_D)$ is an order reversing lattice isomorphism. A covering $(Y_D,f_D,y_D)$ corresponding to $D = \langle e\rangle \subset \pi_1(X,x_0)$ is called a (the) universal covering of $X$ and $Y_{\langle e\rangle}$ is simply connected.

<u>4.3. Lifting conformal structures</u>. If $f : Y \to X$ is a covering of surfaces and $X$ is a Riemann surface then there is a unique conformal structure on $Y$ so that $f$ is holomorphic. For if $x \in X$ and $\Delta$ is a small simply connected open neighborhood which is mapped by $h$ biholomorphically onto $\{|z| < 1\}$ then on $f^{-1}(\Delta) = \bigcup \Delta_i$, $h \circ f$ restricted to each $\Delta_i$ gives the conformal structure for $Y$. If $\tilde{X}$ is the universal covering of $X$ then $\tilde{X}$ is a simply connected Riemann surface.

<u>Uniformization Theorem</u>. Every simply connected Riemann surface is conformally equivalent to $\mathbb{P}^1$, $\mathbb{C}$, or $\mathbb{D}$ (the unit disk).

<u>4.4. Cover transformations</u>.

<u>Notational convention</u> for Sections 4.4, 4.5, 4.6, and 4.7.

We will write all functional notation with the argument on the left; e.g. $x \mapsto (x)f$. Then the composition of $f$ followed by $g$ will be written

$f \circ g$ or $fg$. This notational convention will help distinguish between isomorphisms and anti-isomorphisms which occur naturally in discussing cover transformations.

Suppose $f : Y \to X$ is a covering. A homeomorphism $g : Y \to Y$ will be called a <u>cover transformation</u> if $g \circ f = f$. Under composition the set of cover transformations is a group denoted $G(Y/X)$ and is called the <u>Galois group</u> of the covering. If $X$ and $Y$ are Riemann surfaces and $f$ is holomorphic the cover transformations are biholomorphic.

<u>Lemma 4.4</u>. A cover transformation with a fixed point is the identity.

<u>Proof</u>. Suppose $g \in G(Y/X)$ and $(y_0)g = y_0$. If $y_1$ is any point on $Y$ let $y_2 = (y_1)g$. Let $\tilde{\gamma}$ be a path from $y_0$ to $y_1$. Then $\tilde{\gamma} \circ g$ is a path from $y_0$ to $y_2$. But $(\tilde{\gamma})f = (\tilde{\gamma} \circ g)f = \gamma$, a path in $X$ starting at $x_0$. $\tilde{\gamma}$ and $\tilde{\gamma} \circ g$ are lifts of $\gamma$ starting at $y_0$. By the uniqueness of lifts $\tilde{\gamma} = \tilde{\gamma} \circ g$ and so $y_1 = y_2$.

q.e.d.

<u>Theorem 4.4</u>. Let $(Y, f, y_0)$ be a covering of $(X, x_0)$. Let $D = f_*\big(\pi_1(Y, y_0)\big) \subset \pi_1(X, x_0)$. Then $G(Y/X) \cong N(D)/D$ where $N(D)$ is the normalizer of $D$ in $\pi_1(X, x_0)$.

<u>Proof</u>. Fix $g \in G(Y/X)$. Let $\tilde{\sigma}$ be a path in $Y$ connecting $y_0$ to $(y_0)g$. Let $\sigma = \tilde{\sigma} \circ f$. Now the covering $\big(Y, f, (y_0)g\big)$ is equivalent to $(Y, f, y_0)$. Since $g \circ f = f$ we see that $f_*$ maps both $\pi_1(Y, y_0)$ and $\pi_1\big(Y, (y_0)g\big)$ isomorphically onto $D$. But

$$\pi_1\big(Y, (y_0)g\big) = \tilde{\sigma}^{-1}\, \pi_1(Y, y_0)\, \tilde{\sigma}.$$

Thus

$$D = \sigma^{-1} D \sigma.$$

Now consider the map from $G(Y/X) \to N(D)/D$

$$g \to D\sigma_g$$

where $\tilde{\sigma}_g$ connects $y_0$ to $(y_0)g$. We must check several things about this map.

(i) The map is well defined. If $\tilde{\tau}_g$ connects $y_0$ to $(y_0)g$ then $\tilde{\tau}_g = \tilde{\tau}_g \tilde{\sigma}_g^{-1} \tilde{\sigma}_g$. But $\tilde{\tau}_g \tilde{\sigma}_g^{-1}$ is a closed path starting at $y_0$ and so maps onto $D$ by $f_*$. Therefore $\tau_g \in D\sigma_g$.

(ii) The map is injective. If $D\sigma_g = D\sigma_h$ then $\sigma_h \sigma_g^{-1} \in D$.

$\sigma_h \tilde{\sigma}_g^{-1}$ is a closed path on $Y$. Thus $(1)\tilde{\sigma}_h = (1)\tilde{\sigma}_g$ or $(y_0)h = (y_0)g$. Consequently $h = g$.

(iii) The map is surjective. If $\sigma^{-1} D \sigma = D$ let $\tilde{\sigma}$ be the lift of $\sigma$ starting at $y_0$. Let $y_1 = (1)\tilde{\sigma}$. Then $\pi_1(Y, y_1) = \tilde{\sigma}^{-1} \pi_1(Y, y_0) \tilde{\sigma}$. Since $f_*$ maps both of these groups onto $D$ we see that $(Y, f, y_0)$ is equivalent to $(Y, f, y_1)$. The homeomorphism $g : Y \to Y$ so that $y_1 = (y_0)g$ is a cover transformation and $\sigma = \sigma_g$.

(iv) The map is an anti-isomorphism. A path connecting $y_0$ to $(y_0)gh$ is seen to be $\tilde{\sigma}_h(\tilde{\sigma}_g)h$ which projects by $f$ onto $\sigma_h \sigma_g$. Thus $\sigma_{gh} = \sigma_h \sigma_g$.

q.e.d.

<u>Remark</u>. If $\varphi : G \to H$ is an anti-isomorphism then $\psi : G \to H$, $(g)\psi = (g^{-1})\varphi$ is an isomorphism.

<u>4.5. Some remarks on permutation groups.</u>

Let $\Sigma$ be a set and let $\mathcal{S}(\Sigma)$ be the group of permutations of $\Sigma$. If $\Sigma = \{1, 2, \cdots, n\} \subset \mathbb{Z}$ then $\mathcal{S}(\Sigma)$ will be denoted $\mathcal{S}_n$. A subgroup $H \subset \mathcal{S}(\Sigma)$ is said to be <u>transitive</u> if for all $x, y \in \Sigma$ there is a $h \in H$ so that $(x)h = y$. A subgroup $H$ is said to be <u>regular</u> if the only element in $H$ which fixes any element of $\Sigma$ is the identity. If $H$ is regular and $h \in H$, $h \neq e$, then when we write $h$ as a product of disjoint cycles, all cycles have the same length.

<u>Examples</u>. The right and left Cayley representations. If $G$ is a group the right Cayley representation $T : G \to \mathcal{S}(G)$ is given by $g \to T_g$ where $(h)T_g = hg$. Then $T$ is an injective homomorphism and $(G)T \subset \mathcal{S}(G)$ is a regular transitive subgroup. The left Cayley representation $\tau : G \to S(G)$ is given by $g \to \tau_g$ where $(h)\tau_g = gh$. $\tau_g$ is an injective anti-homomorphism and again $(G)\tau$ is a regular transitive subgroup of $\mathcal{S}(G)$. The two representations are the same if and only if $G$ is abelian.

<u>4.6</u>. Suppose $(Y, f, y_0) \to (X, x_0)$ is a covering of Riemann surfaces. Then $G(Y/X)$ acting on the fiber of $f$ over $x_0$ is a regular group of permutations since only the identity cover transformation has a fixed point. Let this fiber, $(x_0)f^{-1}$ be denoted by $I$.

From the proof of Theorem 4.4 we see that $G(Y/X)\,|\,I$ is transitive if and only if $D$ is normal in $\pi_1(X,x_0)$ $\left(D = \left(\pi_1(Y,y_0)\right)f_*\right)$. In this case the covering is said to be Galois or normal. Assuming the covering is Galois, for $x \in X$, $y \in (x)f^{-1}$ we have $(x)f^{-1} = \{(y)g \mid g \in G(Y/X)\}$. Denote this set, the $G(Y/X)$-orbit of $y$, by $[y]$. Let $\{[y] \mid y \in Y\}$ be denoted by $Y/G(Y/X)$. Since each $G(Y/X)$-orbit $[y]$ can be identified with $(y)f$, $Y/G(Y/X)$ can be given the structure of a Riemann surface and $Y/G(Y/X) \cong X$.

Lemma 4.6. Suppose $(Y,f,y_0)$ is a Galois covering of $(X,x_0)$ and $H$ is a subgroup of $G(Y/X)$. Then there is a factoring of the covering $f: Y \to X$, $Y \to Z \to X$, where $Y \to Z$ is a Galois covering and $G(Y/Z) = H$. $H$ is normal in $G(Y/X)$ if and only if the covering $Z \to X$ is a Galois covering. In this case

$$G(Z/X) \cong G(Y/X)\,/\,G(Y/Z).$$

Proof: Let $D$ be the image of $\pi_1(Y,y_0)$ under $f_*$. Then $D$ is normal in $\pi_1(X,x_0)$ and $\pi_1(X,x_0)/D \cong G(Y/X)$. Let $\varphi$ be the quotient map of $\pi_1(X,x_0)$ onto $G(Y/X)$. Then $D \subset (H)\varphi^{-1} \subset \pi_1(X,x_0)$. Let $Z$ be the Riemann surface which is the covering of $X$ corresponding to $(H)\varphi^{-1}$. Since $D$ is normal in $(H)\varphi^{-1}$ the covering $Y \to Z$ is Galois with Galois group isomorphic to $(H)\varphi^{-1}/D$, which is isomorphic to $H$.

If $H$ is normal in $G(Y/X)$ then $(H)\varphi^{-1}$ is normal in $\pi_1(X,x_0)$ and $G(Z/X) \cong \pi_1(X,x_0)/(H)\varphi^{-1} \cong G(Y/X)/H$.

q.e.d.

If $D = \langle e \rangle$ then $\tilde{X}$, the universal covering surface of $X$, is a Galois cover with $G(\tilde{X}/X) \cong \pi_1(X,x_0)$. If $\tilde{X}$ is isomorphic to the unit disk $\mathbb{D}$, then $G(\tilde{X}/X)$ is a group of fixed-point-free conformal self-maps of $\mathbb{D}$ (fractional linear transformations).

4.7. The monodromy group of the cover $(Y,f,y_0) \to (X,x_0)$ [25, p. 198].

Let $I = \{x_0\}f^{-1} = \{y_0, y_1, y_2, \cdots\}$. Suppose $[\gamma] \in \pi_1(X,x_0)$. (Hereafter we will write $\gamma$ instead of $[\gamma]$ since all statements will depend only on the homotopy class of $\gamma$.) We obtain a permutation $(\gamma)\mu$ in $\mathcal{S}(I)$ as follows. Let $\tilde{\gamma}_i$ be the lift of $\gamma$ starting at $y_i$. Then define $(y_i)(\gamma)\mu = (1)\,\tilde{\gamma}_i$.

(i) By the monodromy theorem $\mu$ is well defined as a function from $\pi_1(X,x_0)$ into $\mathcal{S}(I)$.
(ii) $\mu$ is seen to be a homomorphism.
(iii) Since $Y$ is connected the image of $\mu$ is a transitive subgroup of $\mathcal{S}(I)$. It is called the <u>monodromy group</u> of the covering and will be denoted $M(Y/X)$.
(iv) $\left(\pi_1(Y,y_0)\right)f_* = \left\{\gamma \in \pi_1(X,x_0) \,\middle|\, (y_0)(\gamma)\mu = y_0\right\}$.
(v) $\left(\pi_1(Y,y_j)\right)f_* = \left\{\gamma \,\middle|\, (y_j)(\gamma)\mu = y_j\right\}$. Call this group $D_j$.
(vi) Let $\gamma_j$ be a path whose lift, $\tilde{\gamma}_j$, to $Y$ starting at $y_0$ ends at $y_j$. Then $\tilde{\gamma}_j^{-1}\pi_1(Y,y_0)\tilde{\gamma}_j = \pi_1(Y,y_j)$. Consequently $\gamma_j^{-1}D_0\gamma_j = D_j$.
(vii) $\pi_1(X,x_0) = D_0\gamma_0 \cup D_0\gamma_1 \cup D_0\gamma_2 \cup \cdots$, a right coset decomposition of $\pi_1(X,x_0)$. $D_0\gamma_j$ is the set of paths which when lifted to start at $y_0$ end at $y_j$.

(viii) $\ker\mu = \bigcap_j D_j = \bigcap_{\sigma\in\pi_1(X,x_0)} \sigma^{-1}D_0\sigma$. This is the largest normal subgroup of $\pi_1(X,x_0)$ lying in $D_0$. Now if the cover $Y \to X$ is Galois then $D_0 = \ker\mu$ and $M(Y/X)$ is a regular subgroup of $\mathcal{S}(I)$.
(ix) Thus if $Y \to X$ is Galois we have two regular transitive subgroups of $\mathcal{S}(I)$ namely $M(Y/X)$ and $G(Y/X)\,\big|\,I$. Both groups are isomorphic to $\pi_1(X,x_0)/\left(\pi_1(Y,y_0)\right)f_*$.
(x) How are these two subgroups of $\mathcal{S}(I)$ related?
Let $G = \pi_1(X,x_0)/D_0 = \{D_0\gamma_0, D_0\gamma_1, \cdots\}$. Identify $y_j$ with $D\gamma_j$. If $\mu$ is the monodromy map then $(y_j)(\gamma)\mu = (1)(\tilde{\gamma}_j\gamma)$. That is, $(\gamma)\mu$ acting on $D\gamma_j$ is $D\gamma_j\gamma = D\gamma_j D\gamma$. The monodromy map is thus the right Cayley representation. If $g_j \in G(Y/X)$ where $(y_0)g_j = y_j$ then $(y_i)g_j = (1)\tilde{\gamma}_j\tilde{\gamma}_i$. (See the proof of Theorem 4.4.) That is, $g_j$ acting on $D\gamma_i$ is $D\gamma_j\gamma_i = D\gamma_j D\gamma_i$. Thus $G(Y/X)\,\big|\,I$ is the left Cayley representation.

<u>Theorem 4.7</u>. Suppose $(Y,f,y_0) \to (X,x_0)$ is a Galois covering. Then the

Galois group acting on $(x_0)f^{-1}$ $(=I)$ and the monodromy group are regular transitive subgroups of $\mathcal{S}(I)$. If we identify $I$ with the right cosets $\{D\gamma_i\}$ of $D$ in $\pi_1(X,x_0)$ as above $\left(D=\left(\pi_1(Y,y_0)\right)f_*\right)$, then $M(Y/X)$ acting on $I$ is the right Cayley representation and $G(Y/X)$ is the left Cayley representation.

Corollary. The two representatons are the same if and only if $G(Y/X)$ is abelian.

4.8. We now abandon the notational convention adopted at the beginning of Section 4.4.

Theorem. (Hurwitz). Let $G$ be a finite group. Then there exists a compact Riemann surface $Y$ which admits a group of automorphisms isomorphic to $G$.

Proof. Let $g_1,\cdots,g_s$ be generators for $G$. Let $X$ be a Riemann surface of genus $s$. Let $A_1,\cdots,A_s, B_1,\cdots,B_s$ be a canonical basis for $\pi_1(X,x_0)$. Then $\prod[A_i,B_i]=e$, and this is the only defining relation. Then $\mu:\pi_1(X,x_0)\to G$ can be defined where $\mu(A_i)=g_i$ and $\mu(B_i)=e$ and $\mu$ is a homomorphism from $\pi_1(X,x_0)$ onto $G$. The covering Riemann surface $Y$ corresponding to the kernal of $\mu$ has $G(Y/X)\cong G$.

q.e.d.

4.9 Galois (or normal) closure.

Let $(Y,f,y_0)\to(X,x_0)$ be a covering of surfaces where $D=f_*\left(\pi_1(Y,y_0)\right)$. Let $N=\bigcap_{\sigma\in\pi_1(X,x_0)}\sigma^{-1}D\sigma$ which is the kernel of the monodromy map. The covering $Y_N$ corresponding to $N$ is called the Galois (or normal) closure for the covering $Y\to X$. Note that $G(Y_N/X)=M(Y/X)$.

4.10. Branched coverings.

The coverings of surfaces we have considered so far in this chapter will often be called smooth coverings in contradistinction to branched coverings. The latter will be continuous maps between surfaces which locally are equivalent to $z\to z^n$. An example of a branched covering is a non-constant holomorphic map between two compact Riemann surfaces.

Consider the map from $f:\mathbb{D}-\{0\}\to\mathbb{D}-\{0\}$, $f(z)=z^n$, a smooth covering of Riemann surfaces. Any finite sheeted smooth covering of

$\mathbb{D} - \{0\}$ is of this type. f can obviously be extended to a holomorphic map of $\mathbb{D} \to \mathbb{D}$ by sending $0 \to 0$. f so extended is now a branched covering.

(i) Let X be a Riemann surface. Let B be a discrete set in X. (If X is compact B is finite.) Let $X^* = X - B$ and let $f : Y^* \to X^*$ be a finite sheeted smooth covering. If $b_0 \in B$, let $\Delta \subset X$ be a parametric disk centered at $b_0$ so that $\Delta \cap B = \{b_0\}$. $f^{-1}(\Delta - \{b_0\})$ is a collection of finite sheeted coverings of $\Delta - \{b_0\}$. To each component of $f^{-1}(\Delta - \{b_0\})$ we add a point to $Y^*$ above $b_0$ and extend f to these points to map onto $b_0$. We do this for each point $b_0 \in B$. Let $Y = Y^* \cup$ {points added above B} and $f^e : Y \to X$ is f extended. Then $f^e : Y \to X$ is a finite-sheeted holomorphic map branched (if at all) only above B.

Again consider $b_0 \in \Delta$, $b_0 \in B$ as above. A path $\gamma \in \pi_1(X^*, x_0)$, $x_0 \in X^*$, will be said to "circle" $b_0$ if $\gamma = \sigma\tau\sigma^{-1}$ where $\sigma$ is a path from $x_0$ to $x_1 \in \Delta - \{b_0\}$ and $\tau$ is a path in $\Delta - \{b_0\}$ beginning and ending at $x_1$ and winding once (positively) around $b_0$. If $\mu$ is the monodromy representation for the cover $Y^* \to X^*$ then $\mu(\gamma)$ is a permutation, each cycle of which corresponds to a component of $f^{-1}(\Delta - \{b_0\})$; that is, each cyle of $\mu(\gamma)$ corresponds to a point of Y above $b_0$. The length of a cycle in $\mu(\gamma)$ will be the multiplicity of $f^e$ at the corresponding point of Y.

(ii) Also any cover transformation g of $G(Y^*/X^*)$ extends to a biholomorphic map $g^e$ of $Y \to Y$, and since $f \circ g = f$ we have $f^e \circ g^e = f^e$. Such a map of $Y \to Y$ will be called a cover transformation of Y and the group of such cover transformations will be denoted $G(Y/X)$.

(iii) Let X be a compact Riemann surface, B a finite set $\{b_1, \cdots, b_s\}$, $X^* = X - B$. Let D be a normal subgroup of $\pi_1(X^*, x_0)$, $x_0 \in X^*$, and let $Y^*$ be the corresponding Galois cover of $X^*$, $f : Y^* \to X^*$. We then extend $Y^*$ to Y, f. to $f^e$, and $G(Y^*/X^*)$ to $G(Y/X)$. Then $G(Y/X)$ is isomorphic to $\pi_1(X^*, x_0)/D$. Now for each $b_j \in B_0$

let $\gamma_j$ be a path circling $b_j$. Then $\mu(\gamma_j)$ is a regular permutation of length, say, $v_j$. The ramification of $f^e$ above $b_j$ is then seen to be $(n/\nu_j)(\nu_j - 1)$ where $n$ is the number of sheets in the covering. We obtain <u>The Riemann-Hurwitz formula for Galois coverings</u>

$$2p_Y - 2 = n(2p_X - 2) + \sum_{j=1}^{s} n(1 - \frac{1}{\nu_j}) .$$

<u>Lemma 4.10</u>. Let $G$ be a finite group of automorphisms on a Riemann surface $X$. Suppose there is an $x_0 \in X$ so that $g(x_0) = x_0$ for all $g \in G$. Then $G$ is a cyclic group.

<u>Proof</u>. Let $\Delta$ be a parametric disk centered at $x_0$. Let $\Delta_0$ be the component of $\bigcap_{g \in G} g\Delta$ containing $x_0$. Then $\Delta_0$ is simply connected and invariant under $G$. By the uniformization theorem we may assume that $G$ acting on $\Delta_0$ is a finite group of fractional linear transformation of $\mathbb{D} \to \mathbb{D}$ which leaves zero fixed. It now follows that $G$ is cyclic.

(iv) Let $Y$ be a Riemann surface with a finite group automorphisms $G$. Let $Y/G$ be the set of orbits of $G$. If the only element of $G$ with a fixed point is the identity then $Y/G$ is made a Riemann surface $X$ by giving it the quotient topology and requiring that the natural map $f : Y \to X$ be holomorphic. In fact, $f$ is a smooth covering.

Suppose $G$ admits non-identity elements with fixed points. For $y \in Y$ let $St(y) = \{g \in G \mid g(y) = y\}$, the stabilizer of $y$. If $h \in G$ then $St(h(y)) = h\, St(y) h^{-1}$. Thus the fixed points of non-identity elements of $G$ are orbits of $G$.

Let $Y^* = \{y \mid St(y) = e\}$. Then $G - \{e\}$ is fixed-point-free on $Y^*$. Let $X^* = Y^*/G$. By the previous discussion $f : Y^* \to X^*$ can be extended to $f^e : Y \to X$ where we add to $X$ (points corresponding to) the orbits of points in $Y$ with stabilizers larger than $\langle e \rangle$. $f^e$ is now holomorphic and branched precisely over $X - X^*$.

We now summarize this discussion in a theorem.

<u>Theorem 4.10</u>. If $G$ is a finite group of automorphisms on a Riemann surface $Y$, then the orbit space $Y/G$ is naturally a Riemann surface so that the projection $f : Y \to Y/G$ is a holomorphic map. The branch-

ing of $f$ occurs precisely at the fixed points of non-identity elements of $G$. The number of sheets in the covering $Y \to Y/G$ is the order of $G$.

(With a slight modification of the above discussion one sees that this theorem is true for countable groups of automorphisms, $G$.)

4.11. We saw in Lemma 4.6 that if $Y \to Z$ is a Galois covering and $H$ is normal in $G(Y/Z)$ then there exist a factoring $Y \to X \to Z$ where both intermediate coverings are normal. We want to reverse this procedure. If $Y \to X$ and $X \to Z$ are Galois coverings, when do we know that $Y \to Z$ is Galois? This then is the Riemann surface (or function field) version of a basic problem in group extensions.

Theorem 4.11. (A. M. Macbeath [18]). Let $X$ be a surface. Let $D$ be a normal subgroup of $\pi_1(X,x_0)$ and let $(Y,f,y_0) \to (X,x_0)$ be the corresponding smooth Galois covering. Let $\varphi$ be a homeomorphism of $X$ (possibly with fixed points). Let $\alpha$ be a path in $X$ joining $x_0$ to $\varphi(x_0)$. Define $\varphi_* : \pi_1(X,x_0) \to \pi_1(X,x_0)$ by

$$\varphi_*(\gamma) = \alpha\,\varphi(\gamma)\,\alpha^{-1} .$$

Suppose $\varphi_*(D) = D$.

(Since $D$ is normal this criterion is independent of $\alpha$.)

Let $y_1$ be any point above $\varphi(x_0)$.

Then there exists a homeomorphism $\tilde{\varphi}$ of $Y$ onto $Y$ so that $\tilde{\varphi}(y_0) = y_1$, and $f \circ \tilde{\varphi} = \varphi \circ f$. That is, $\tilde{\varphi}$ preserves the fibers of $f$.

Consequently $\tilde{\varphi}$ normalizes $G(Y/X)$.

Proof. Let $\tilde{\alpha}$ be a path in $Y$ connecting $y_0$ to $y_1$. Let $\alpha = f(\tilde{\alpha})$, connecting $x_0$ to $\varphi(x_0)$. If $y \in Y$ let $\tilde{\sigma}$ connect $y_0$ to $y$. If $\sigma = f(\tilde{\sigma})$ let $x = \sigma(1)$. $\varphi(\sigma)$ connects $\varphi(x_0)$ to $\varphi(x)$. Let $\tilde{s}$ be the lift of $\varphi(\sigma)$ starting at $y_1$. Define $\tilde{\varphi}(y) = \tilde{s}(1)$.

To show that $\tilde{\varphi}$ is well defined, suppose $\tilde{\tau}$ connects $y_0$ to $y$. Then $\tau = f(\tilde{\tau})$ connects $x_0$ to $x$. Now $\tilde{\sigma}\tilde{\tau}^{-1}$ is a closed path in $Y$ so $\sigma\tau^{-1}$ is a closed path in $D$. Let $\tilde{t}$ be the lift of $\varphi(\tau)$ starting at $y_1$. Now $\varphi_*(\sigma\tau^{-1})$ is in $D$. Therefore, the lifts of $\alpha\,\varphi(\sigma)$ and $\alpha\,\varphi(\tau)$ have the same endpoint in $Y$; that is, $\tilde{t}(1) = \tilde{s}(1)$. This shows that $\tilde{\varphi}$ is well defined.

That $\tilde{\varphi}$ is locally $f^{-1} \circ \varphi \circ f$ shows that $\tilde{\varphi}$ is a local homeomorphism. That $\tilde{\varphi}$ has an inverse sending $y_1$ to $y_0$ is seen by inverting the steps in the proof.

4.12. If $\varphi$ is an element in a group, let $o(\varphi)$ denote its order.

If $o(\varphi)$ is finite in the previous theorem, it will always be the case that $o(\varphi)$ divides $o(\tilde{\varphi})$, but equality is not necessary. We now give a simple criterion for $o(\varphi) = o(\tilde{\varphi})$. (In the language of group extensions this means that the extension

$$0 \to K \to \langle K, \tilde{\varphi} \rangle \to \langle \varphi \rangle \to 0$$

splits, where $K = G(Y/X)$).

Corollary 4.12. Suppose we have the hypotheses of Theorem 4.11 where $X$ and $Y$ are Riemann surfaces and $\varphi$ and $\tilde{\varphi}$ are biholomorphic. Suppose $\varphi$ has a fixed point. Then $\tilde{\varphi}$ can be choosen so that $o(\varphi) = o(\tilde{\varphi})$.

Proof. Let $x_0$ be a fixed point of $\varphi$. Suppose $f(y_0) = x_0$ Then we can find $\tilde{\varphi}$ so that $\tilde{\varphi}(y_0) = y_0$. Now $(\tilde{\varphi})^{o(\varphi)}$ is in $G(Y/X)$ and has a fixed point $y_0$; consequently $(\tilde{\varphi})^{o(\varphi)} = e$. Therefore $o(\tilde{\varphi})$ divides $o(\varphi)$.

q.e.d.

4.13. When $G(Y/X)$ is abelian the criterion for lifting automorphisms can be stated in terms of the first homology group of $X$, $H_1(X,Z) \cong \pi_1(X,x_0)/[\pi_1(X,x_0), \pi_1(X,x_0)]$.

Theorem 4.13. Suppose $f: Y \to X$ is a smooth Galois covering of surfaces where $G(Y/X)$ is abelian. Let $D$ be the subgroup of $\pi_1(X,x_0)$ corresponding to this covering. Let $D'$ be the image of $D$ under the quotient map of $\pi_1(X,x_0)$ onto $H_1(X,Z)$. A homeomorphism $\varphi$ of $X$ acts on $H_1(X,Z)$ in a natural way. If such an aciton of $\varphi$ on $H_1(X,Z)$ leave $D'$ invariant, then $\varphi$ lifts to $Y$ as in Theorem 4.11.

Proof. If $\gamma \in D$ then $\varphi_*(\gamma) \in D \cdot [\pi_1(X,x_0), \pi_1(X,x_0)]$. Therefore $\varphi_*(\gamma) \in D$ since the commutator subgroup is a subgroup of $D$.

Corollary 4.13. Let $Y \to X$ be a smooth Galois cover of Riemann surfaces where $G(Y/X)$ is abelian. Suppose $X$ is hyperelliptic. Then the hyperelliptic involution lifts to $Y$.

Proof. In $H_1(X,Z)$ the hyperelliptic involution sends every cycle into its inverse and, therefore, preserves all subgroups of $H_1(X,Z)$.

4.14. Conjugate automorphism groups.

Lemma 4.14. Let $X$ be a Riemann surface with a finite group of automorphisms $G$. Let $\varphi$ be another automorphism, $\varphi \notin G$. Then

$X/G$ is conformally equivalent to $X/\varphi G \varphi^{-1}$.

Proof. Let $[x]$ be a G-orbit. Then $[\varphi(x)]$ is a $\varphi G \varphi^{-1}$-orbit. For if $y = g(x)$ then $\varphi(y) = \varphi g \varphi^{-1}(\varphi(x))$. Thus the map $[x] \to [\varphi(x)]$ takes $X/G$ onto $X/\varphi G \varphi^{-1}$.

4.15. We now include some consequences of the previous paragraphs which will be of use later.

4.15.1. A Riemann surface of genus 2 does not admit an automorphism of order 7. A Riemann surface of genus 3 does not admit an automorphism of order 5.

4.15.2. A Riemann surface of even genus does not admit an automorphism group isomorphic to $Z_2 \times Z_2 \times Z_2$ where $Z_2$ is the cyclic group of order 2.

4.15.3. Let $Y \to X$ be a Galois covering of compact Riemann surfaces where $G(Y/X)$ is abelian. If the covering is not smooth, then it must be branched above at least 2 points of $X$.

The first three results follow from the Riemann-Hurwitz formula for Galois covers. The fourth follows from the observation that the commutator subgroup of an abelian group is the identity.

## Chapter 5. Automorphisms of Riemann surfaces, I.

5.1. If $X$ is a Riemann surface let $\mathrm{Aut}(X)$ stand for the group of automorphisms of $X$.

We now reproduce, with slight modifications, Hurwitz's proof that a group of automorphisms on a compact Riemann surface of genus greater than one is finite [15].

Lemma 5.1. Let $X$ be a compact Riemann surface of genus $p$. Let $\varphi$ be an automorphism of $X$. Then $\varphi$ has at most $2p + 2$ fixed points.

Proof. Suppose $\varphi$ has $s$ fixed points. Let $f$ be a meromorphic function with precisely $p + 1$ poles at $p + 1$ distinct points, none of which are fixed points for $\varphi$ and so that none of the poles of $f$ and $f \circ \varphi$ coincide. Then $f - f \circ \varphi$ has $2p + 2$ poles and at least $s$ zeros. Therefore $2p + 2 \geq s$.

q.e.d.

Theorem 5.1. (H. A. Schwarz [23]). If $X_p$ is a compact Riemann surface with genus $p \geq 2$, then $\mathrm{Aut}(X_p)$ is a finite group.

Proof. For $p \geq 2$ the total weight of the $3K$ points (higher order Weierstrass points for the tri-canonical series) is $25(p-1)^2 p$. The maximum weight at any point is $p(p+1)/2$. Thus there are at least $25(p-1)^2 p / \left(p(p+1)/2\right)$ distinct $3K$ points. For $p \geq 2$ this number is greater than $2p + 2$.

Any automorphism permutes the $3K$-points. We thus have a map from $\mathrm{Aut}(X_p) \to \mathcal{S}(3K\text{-points})$. The kernel of this map fixes the $3K$-points, and so by Lemma 5.1 the kernel is the identity. Thus $\mathrm{Aut}(X_p)$ is isomorphic to a subgroup fo a finite permutation group.

q.e.d.

Definition. For $p \geq 2$ let $N(p)$ be the largest order of an automorphism group on a Riemann surface of genus $p$.

Many applications of what follows will give information about the nature of this function, $N(p)$.

5.2. Let $f : Y \to \mathbb{P}^1$ be a Galois covering of $n$-sheets, $Y$ compact. Suppose the ramification occurs above $B = \{z_1, \ldots, z_s\} \subset \mathbb{P}^1$. Let $\gamma_i$ be a path circling $z_i$ so that

$\pi_1(\mathbb{P}^1 - B, z_0) = \langle \gamma_1, \gamma_2, \ldots, \gamma_s \mid \gamma_1 \gamma_2 \cdots \gamma_s = e \rangle$. Let $\nu_i$ be the ramification of a single point over $z_i$. Then $\gamma_i^{\nu_i} = e$. If

$\mu : \pi_1(\mathbb{P}^1 - B, z_0) \to \mathcal{S}_n$ is the monodromy map then $M(Y/\mathbb{P}^1)$ is generated by $\mu(\gamma_1),\cdots,\mu(\gamma_s)$. If $\mu(\gamma_i) = a_i \in \mathcal{S}_n$ then the $a_i$'s satisfy the following: $a_i^{\nu_i} = e$, $i = 1,2,\cdots,s$, and $a_1 a_2 \cdots a_s = e$. These generators obviously can be expected to satisfy other relations.

Definition. A finite group generated by elements $a_1,\cdots,a_s$ so that $a_1^{\nu_1} = \cdots = a_s^{\nu_s} = a_1 a_2 \cdots a_s = e$ will be denoted $(\nu_1,\cdots,\nu_s)$.

In the discussion preceeding the definition both $G(Y/\mathbb{P}^1)$ and $M(Y/\mathbb{P}^1)$ are $(\nu_1,\cdots,\nu_s)$ groups since the two groups are isomorphic. If $p$ is the genus of $Y$ then the Riemann-Hurwitz formula is:

$$(*) \quad 2p - 2 = -2n + n\sum_{k=1}^{s}\left(1 - \frac{1}{\nu_k}\right) = n\left(s - 2 - \sum_{k=1}^{s}\frac{1}{\nu_k}\right).$$

When $G(Y/\mathbb{P}^1)$ is abelian, in the identification of $G(Y/\mathbb{P}^1)$ with $M(Y/\mathbb{P}^1)$ the permutation $\mu(\gamma_i)$ is equal to the stabilizer of one (and therefore all) of the points in $Y$ above $z_i$ in $\mathbb{P}^1$. This will, in fact, be the case for arbitrary $G(Y/\mathbb{P}^1)$ whenever $\mu(\gamma_i)$ is central in $M(Y/\mathbb{P}^1)$. The observations of this paragraph are valid for any Galois cover $Y \to X$ and do not depend on the genus of $X$ being zero. All these remarks follow from Theorem 4.7.

5.3. Groups of genus zero.

In 5.3 assume that $p = 0$, that is $Y = \mathbb{P}^1$.

Definition. A finite group fo automorphism of $\mathbb{P}^1$ will be called a group of genus zero.

In the fomula $(*)$ of 5.2 we see that $s - 2 - \sum_{k=1}^{s}\frac{1}{\nu_k}$ must be negative, equal to $-\frac{2}{n}$, in fact. Since $\nu_k \geq 2$ we see that $s$ must be less than or equal to $3$.

If $s = 2$ $G(\mathbb{P}^1/\mathbb{P}^1)$ is an $(n,n)$, a cyclic group of order $n$.

If $s = 3$ we list the possibilities in the following table:

| $G(\mathbb{P}^1/\mathbb{P}^1)$ | order | description |
|---|---|---|
| (2,2,n) | 2n | Dihedral group $D_n$ |
| (2,3,3) | 12 | Alternating group $A_4$ |
| (2,3,4) | 24 | Symmetric group $\mathcal{S}_4$ |
| (2,3,5) | 60 | Alternating group $A_5$ |

This allows us to completely describe hyperelliptic automorphism groups.
Theorem 5.3. Let $X_p$ be a compact hyperelliptic Riemann surface, $p \geq 2$. Let $H$ be the hyperelliptic involution. Then the subgroup of order 2, $\langle H \rangle$, is normal (and central) in $\mathrm{Aut}(X_p)$ and $\mathrm{Aut}(X_p)/\langle H \rangle$ is a group of genus zero.

Proof. Since the $g^1_2$ is unique, the hyperelliptic involution is unique. But for $g \in \mathrm{Aut}(X_p)$, $g^{-1}Hg$ is also a hyperelliptic involution (Lemma 4.14). Consequently $g^{-1}Hg = H$. The result now follows from Lemma 4.6.

5.4. Groups of genus one.

If $X_1$ is a Riemann surface of genus one, any finite group of $X_1$ will be called a group of genus one provided the quotient is $\mathbb{P}^1$. Since $\pi_1(X_1, \cdot)$ is isomorphic to $Z \times Z$, we can find a fixed point free finite group of automorphism isomorphic to $Z_m \times Z_n$ for any two positive integers $m,n$.

For a Galois cover $X_1 \to \mathbb{P}^1$ the Riemann-Hurwitz formula is

$$\left(s - 2 - \sum_{k=1}^{s} \frac{1}{\nu_k}\right) = 0$$

with no obvious condition on $n$. The four possibilities are

| | |
|---|---|
| s = 4 | (2,2,2,2) |
| s = 3 | (2,3,6) |
| | (2,4,4) |
| | (3,3,3) |

In analogy with hyperelliptic Riemann surfaces we define a Riemann surface to be elliptic-hyperelliptic if it is a 2-sheeted branched covering of a Riemann surface of genus one. An involution whose orbits are the fibers of this 2-sheeted covering will be called an elliptic-hyperelliptic involution. Later in this chapter we will show that an elliptic-hyperelliptic Riemann surface of genus greater than 5 has a unique elliptic-hyperelliptic involution. (This also follows from the Castelnuovo-Severi inequality.) Then we can repeat the proof of Theorem 5.3 to conclude that for genus greater than 5, an elliptic-hyperelliptic automorphism group has a normal (central) subgroup of order two, and the quotient group is a group of genus one.

5.5. Maximal automorphism groups for $p \geq 2$.

We now reproduce Hurwitz's proof that for $p \geq 2$

$$N(p) \leq 84(p-1) \qquad [15].$$

Lemma 5.5. Let $Y \to X$ be a Galois covering of $n$ sheets of Riemann surfaces. Assume $p_y$, the genus of $Y$, is greater than one. If $p_x$, the genus of $X$, is positive then

$$o\big(G(Y/X)\big) \leq 4(p_y - 1).$$

Proof. The Riemann-Hurwitz formula for the covering is

$$2p_y - 2 = n(2p_x - 2) + n \sum_{i=1}^{s} \left(1 - \frac{1}{\nu_i}\right).$$

If $p_x \geq 2$ then $2p_y - 2 \geq n(2p_x - 2) \geq 2n$, or $n \leq p_y - 1$.

If $p_x = 1$ then $2p_y - 2 = n \sum_{i=1}^{s} \left(1 - \frac{1}{\nu_i}\right) \geq \frac{1}{2} n$.

q.e.d.

Thus to obtain large automorphism groups we must have $p_x = 0$.

Then $2p_y - 2 = n\left(s - 2 - \sum_{i=1}^{s} \frac{1}{\nu_i}\right)$.

If $s \geq 5$ then $2p_y - 2 \geq n\left(3 - \frac{5}{2}\right) = \frac{1}{2} n$.

If $s = 4$ then $2p_y - 2 = n\left(2 - \sum_{i=1}^{4} \frac{1}{\nu_i}\right)$.

Since $p_y > 1$, not all the $\nu_i$ can equal $2$. The order of $G(Y/X)$, $n$, attains its maximum when $(\nu_1, \nu_2, \nu_3, \nu_4) = (2,2,2,3)$ and $n = 12(p_y - 1)$.

If $s = 3$ then $2p_y - 2 = n\left(1 - \sum_{i=1}^{3} \frac{1}{\nu_i}\right)$.

In order to assure that $p_y > 1$, we must exclude those values of $(\nu_1, \nu_2, \nu_3)$ which occur for groups of genus zero and one. Thus

we choose $(\nu_1, \nu_2, \nu_3)$ so that $\left(1 - \sum \frac{1}{\nu}\right)$ is positive but as small as possible. The appropriate values are $(2,3,7)$ in which case $n = 84(p_y - 1)$.

Theorem 5.5. (Hurwitz). For $p \geq 2$, $N(p) \leq 84(p-1)$. If a Riemann surface of genus $p$ admits a group $G$ of automorphisms of order $84(p-1)$ then $G$ is a $(2,3,7)$ and the quotient map $X_p \to \mathbb{P}^1$ is branched over 3 points of $\mathbb{P}^1$ above which the multiplicities of the branch points are $2,3$, and $7$.

Conversely, if $G$ is an abstract $(2,3,7)$, then there exists a Riemann surface $X_p$ admitting an automorphism group isomorphic to $G$ of order $84(p-1)$.

Proof. To prove the statement in the second paragraph, let $G$ be an abstract $(2,3,7)$, $G = \langle a, b \rangle$ where $a^2 = b^3 = (ab)^7 = e$. Define $\mu : \pi_1\left(\mathbb{P}^1 - \{0,1,\infty\}, 2\right) \to G$ by $\mu(\gamma_0) = a$, $\mu(\gamma_1) = b$ and $\mu(\gamma_\infty) = (ab)^{-1}$. The kernel of $\mu$ corresponds to a Galois covering $Y^*$ of $\mathbb{P}^* \left(= \mathbb{P}^1 - \{0,1,\infty\}\right)$ with $G(Y^*/\mathbb{P}^*) \cong G$. Filling in the branch points over $0,1$, and $\infty$ gives a branched Galois covering $Y \to \mathbb{P}^1$ and $N(p_y) = 84(p_y - 1)$.

q.e.d.

Thus the problem of finding Riemann surfaces admitting automorphism groups of maximum order is the purely group theoretic problem of finding finite $(2,3,7)$'s.

To show that $N(p) = 84(p-1)$ for an infinite number of $p$'s, one must find an infinite number of $(2,3,7)$'s. This was done by G. A. Miller [21] in 1902, about 10 years after Hurwitz's original work.

5.6. An infinite number of $(2,3,7)$'s.

We shall follow A. M. Macbeath's discussion of $(2,3,7)$'s [18]. We first show the existence of a famous group.

Lemma 5.6. There exist a $(2,3,7)$ of order $168$.

Proof. Let $G = GL(3,2)$, the group of all $3 \times 3$ non-singular matrices over the field $K_2$ of two elements. By counting the number of ordered bases in the 3-dimensional vector space over $K_2$ we see that the order of $G$ is $168 = 7 \cdot 6 \cdot 4$. Now consider $G$ as the group of automorphisms of the elementary abelian group of order 8, $G_8$. Then $G_8 = \langle a,b,c \rangle$ or $G_8 = \{e, a, b, c, ab, bc, ca, abc\}$

$= \{e, g_1, g_2, g_3, g_4, g_5, g_6, g_7\}$

where we relable the group elements $g_i$ in the manner indicated; e.g. $g_5 = bc$. Let $A$ be the automorphism $a \to b$, $b \to a$, $c \to abc$. Using the subscripts of the $g_j$'s we can describe $A$ by the permutation $(12)(37)$. Let $B = (145)(267)$. Then $C = AB = (1673245)$. If we let $\langle A,B \rangle = H$ then $H$ is a $(2,3,7)$ and we want to show that the order of $H$ is 168. Since 3 and 7 divide the order of $H$, we show $H$ contains a subgroup of order 8. $\langle A, B^{-1}AB \rangle$ is such a subgroup since $AB^{-1}AB = (1372)(46)$, that is $\langle A, B^{-1}AB \rangle = D_4$, a $(2,2,4)$.

q.e.d.

With a little more work one shows that $G$ is simple.

Applying Theorem 5.5 gives us Klein's Riemann surface of genus 3 admitting 168 automorphisms.

Theorem 5.7. (Macbeath [18,19]. There exist a sequence $p_n \to \infty$ and $N(p_n) = 84(p_n - 1)$.

Proof. Let $X_3$ be Klein's Riemann surface. Let $H_n$ be the characteristic subgroup of $\pi_1(X_3, x_0)$ containing the commutator subgroup and all $n^{th}$ powers. $H_n$ is the kernel of the surjective homomorphism

$$\pi_1(X_3, x_0) \to Z_n \times Z_n \times Z_n \times Z_n \times Z_n \times Z_n .$$

Since $H_n$ is characteristic every element of $\mathrm{Aut}(X_3)$ leaves $H_n$ invariant. The Riemann surface $Y_n$ corresponding to $H_n$ is a smooth, $n^6$-sheeted Galois covering of $X_3$ and all of $\mathrm{Aut}(X_3)$ lifts to $Y_n$. The genus of $Y_n$ is $2n^6 + 1$, and the order of $\mathrm{Aut}(Y_n)$ is $168n^6$.

q.e.d.

This method of Macbeath's, that is, lifting automorphisms to Galois coverings corresonding to characteristic subgroups of $\pi_1(X, x_0)$ is very useful.

5.7 Smooth cyclic coverings of prime order.

Let $n$ be a prime integer. We need a criterion which assures that an automorphism of order $n$ on a Riemann surface is fixed point free.

Lemma 5.7. Let $S_n$ be a cyclic group of prime order $n$ on a Riemann surface $X_p$. Suppose $p - 1 = sn$ and $n > 2s + 3$. Then the covering

$$X_p \to X_p/S_n (= Y)$$

is smooth.

Proof. Suppose not. The Riemann-Hurwitz formula for the covering is $2p_X - 2 = n(2p_Y - 2) + t(n - 1)$ where $t$ is the number of branch points and $t > 0$. Then

$$t(n - 1) = 2p_X - 2 - n(2p_Y - 2) = 2n(s + 1 - p_Y).$$

Therefore $n - 1 \mid 2(s + 1 - p_Y)$ or $n - 1 \le 2s + 2$. This is the desired contradiction.

q.e.d.

5.8. $N(p) < 84(p - 1)$ infinitely often.

Theorem. (Macbeath [19]). Let $n$ be a prime greater than $84$. Let $p = n + 1$. Then $N(p) < 84(p - 1)$.

Proof. Suppose $X_p$ admits a group $G$ of order $84(p - 1) = 84n$. Let $S_n$ be an n-Sylow subgroup of $G$, a cyclic group of order $n$. Since $n > 84$, $S_n$ is a normal subgroup of $G$ (Sylow's theorem). By Lemma 5.7 $X_p \to X_p/S_n$ is a smooth covering and $X_p/S_n$ is a Riemann surface of genus $2$ admitting a group of automorphisms of order $84$ (Lemma 4.6). But a Riemann surface of genus $2$ does not admit an automorphism of order $7$ (4.15.1).

q.e.d.

We shall return to this function $N(p)$ in Chapter 7.

5.9. Groups with partitions [2].

We now generalize the hyperelliptic situation.

Definition. A finite group $G_0$ is said to admit a partition if there is a collection of subgroups $G_i$, $i = 1,2,\cdots,s$, $s \ge 3$, so that

(1) $G_0 = \bigcup_{i=1}^{s} G_i$ and (2) if $0 < i < j$ then $G_i \cap G_j = \langle e \rangle$. The examples of such groups that will interest us are groups in which every element has order $2$, and dihedral groups.

Theorem 5.9. Let $X$ be a compact Riemann surface of genus $p$. Suppose $X$ admits a finite group of automorphisms, $G_0$, where $G_0$ admits a partition. Let the pertinent subgroups be $G_1, G_2, \cdots, G_s$.

Let the order of $G_i$ be $n_i$, let $X_i = X/G_i$ and let $p_i$ be the genus of $X/G_i$ for $i = 0,1,2,\cdots,s$. Then:

$$(s-1)p + n_0 p_0 = \sum_{i=1}^{s} n_i p_i .$$

Proof. Let the Riemann-Hurwitz formula for $X \to X/G_i$ be

(*) $$2p - 2 = n_i(2p_i - 2) + \mathrm{ram}_i .$$

We show $\sum_{i=1}^{s} \mathrm{ram}_i = \mathrm{ram}_0$. For $x \in X$ the elements in $G_0$ which leave $x$ fixed form a cyclic group whose generator lies in one of the $G_i$'s. It follows that $St(x)$ lies completely in this $G_i$. Consequently, the contribution of $St(x)$ to $\mathrm{ram}_0$ contributes to one and only one of the $\mathrm{ram}_i$'s.

The theorem now follows by summing equation (*) as $i$ runs from 1 to $s$ and comparing to equation (*) with $i = 0$, noting that

$$n_0 = \sum_{i=1}^{s} n_i + 1 - s .$$

q.e.d.

The theorem is of interest because the ramifications do not enter into the formula. For a beautiful generalization of the material in this section see the paper of Ernst Kani "Relations between the genera and between the Hasse-Witt invariants of Galois coverings of curves" Canadian Mathematical Bulletin Vol. 28 (3), 1985, pp. 321-327.

5.10. $Z_2 \times Z_2$.

We first apply Theorem 5.9 to the non-cyclic group of order 4 (four group). There are 3 subgroups of order 2 so in this case $s = 3$. The formula becomes

$$p + 2p_0 = p_1 + p_2 + p_3 .$$

We use this formula to deduce a folk theorem that appears to be due to Enriques [10].

Lemma 5.10. Let $X_3$ be a covering of an $X_2$. Then $X_3$ is hyperelliptic.

Proof. $X_3 \to X_2$ is smooth and 2-sheeted. Let $T$ be the involution in $\mathrm{Aut}(X_3)$ so that $X_3/\langle T\rangle = X_2$. The hyperellipic involution on $X_2$ lifts to an automorphism $S$ on $X_3$ of order two with some fixed points. $S$

normalizes $\langle T\rangle$ ; consequently, $\langle S,T\rangle$ is a four group on $X_3$ with $X_3/\langle S,T\rangle = \mathbb{P}^1$. Apply the formula above with $p = 3$, $p_0 = 0$ and one of the $p_i$'s equal to 2. The other $p_i$'s must be 0 and 1.

q.e.d.

Similarly, one shows that an $X_5$ (or $X_7$) which is a two-sheeted covering of a hyperelliptic $X_3$ (or $X_4$) must be hyperelliptic or elliptic-hyperelliptic.

5.11. $Z_2 \times Z_2 \times Z_2$.

There are 7 subgroups of $Z_2 \times Z_2 \times Z_2$ of order 2. Theorem 5.9 gives

$$3p + 4p_0 = \sum_{i=1}^{7} p_i .$$

Lemma 5.11. A Riemann surface of genus 3 admitting a $Z_2 \times Z_2 \times Z_2$ is hyperelliptic.

Proof. If $X_3$ is not hyperelliptic then all $p_i$'s above $(i \geq 1)$ must be one, whereas the left-hand side is at least 9.

q.e.d.

In fact, by relabeling, if necessary, it can be arranged that $p_1 = 0$, $p_2 = p_3 = p_4 = 1$, and $p_5 = p_6 = p_7 = 2$.

A hyperelliptic Riemann surface cannot admit $Z_2 \times Z_2 \times Z_2 \times Z_2$ as a group of automorphisms because $Z_2 \times Z_2 \times Z_2$ is not a group of genus zero (Theorem 5.3).

5.12. Dihedral groups of automorphisms.

Let $D_n$ be a dihedral group of order $2n$. Let $R$ generate the cyclic subgroup of order $n$. If $V$ is an element of order 2 not in $\langle R\rangle$ then $V_i = R^i V$ $(i = 1,2,\cdots,n)$ are the elements in $D_n$ not in $\langle R\rangle$. If $D_n$ is realized as a group of automorphisms on a Riemann surface $X$ of genus $p$, let $X_R = X/\langle R\rangle$, $X_i = X/\langle V_i\rangle$, $i = 1,\cdots,n$ and let $X_0 = X/D_n$ with genera $p_R, p_1, \cdots, p_n, p_0$ respectively. Theorem 5.9 now gives

$$np + 2np_0 = np_R + 2\sum_{i=1}^{n} p_i .$$

If $n$ is odd then all the groups $\langle V_i\rangle$ are conjugate and so $p_i = p_1$, for $i = 2,\cdots,R$ (Lemma 4.14). Thus we obtain

$$p + 2p_0 = p_R + 2p_1 .$$

If $n$ is even then $\langle V_i \rangle$ is conjugate to $\langle V_j \rangle$ if and only if $i \equiv j \pmod 2$.

Let $p_1$ and $p_2$ be the two genera in this case. Then

$$(*) \qquad p + 2p_0 = p_R + p_1 + p_2 .$$

This last formula holds for $n$ odd or even with $n$ odd implying $p_1 = p_2$.

5.13. Commuting involutions.

By the Castelnuovo-Severi inequality we know that if $p > 4p_0 + 1$ and $X_p$ admits an involution $T$ so that the genus of $X_p/\langle T \rangle$ is $p_0$ then this property makes $T$ unique. If, however, $p \le 4p_0 + 1$ then $T$ need not be unique. The next application is an extension of these ideas.

Lemma 5.13. Let $p_1$ and $p_2$ be non-negative integers. Let $X$ be a Riemann surface of genus $p$ so that

$$2p \ge 3p_1 + 3p_2 + 3 .$$

Let $X$ admit two distinct involutions $S_1$ and $S_2$ so that $X/\langle S_i \rangle$ has genus $p_i$, $i = 1,2$. Then $S_1$ and $S_2$ commute.

Proof. Let $G_0 = \langle S_1, S_2 \rangle$ a dihedral group of order $2n$. Set $S_1 = V_1$ and $S_2 = V_2$. Then

$$p + 2p_0 = p_1 + p_2 + p_R .$$

We wish to show $n = 2$. Suppose $n \ge 3$. The Riemann-Hurwitz formula for $X \to X/\langle R \rangle$ is

$$2p - 2 = n(2p_R - 2) + \text{ram} \ge 3(2p_R - 2)$$

or

$$3p_R \le p + 2 .$$

Since $p_0 \ge 0$ we have

$$3p \le 3p + 6p_0 = 3p_1 + 3p_2 + 3p_R \le 3p_1 + 3p_2 + p + 2 .$$

This contradiction proves the lemma.

q.e.d.

If $p_1 \ne p_2$ the above result can be strengthened since $n \ne 2$ implies $n \ge 4$; for in this case $S_1$ and $S_2$ cannot be conjugate.

## Chapter 6. When are fixed points of automorphisms exceptional in some other sense?

### 6.1. Lewittes' Theorem.

The following theorem is due to J. Lewittes [17].

Theorem 6.1. Let $X_p$ be a Riemann surface of genus $p \geq 2$. Let $T$ be an automorphism with 5 or more fixed points. Then each fixed point is an ordinary Weierstrass point.

This theorem will follow from the following lemma.

Lemma 6.1. Let $f : X_p \to X_q$ be an $n$-sheeted branched covering of Riemann surfaces. Suppose the total ramification of this covering is greater than $4(n-1)$. If $z$ is a branch point of $f$ of multiplicity $n$ then $z$ is an ordinary Weierstrass point.

Proof. We show that $X_p$ admits a meromorphic function whose only pole is at $z$ and whose order is less than or equal to $p$.

Let $y = f(z)$. Let $g$ be a meromorphic function on $X_q$ whose only pole is at $y$ and whose order is less than or equal to $q+1$. The Riemann-Hurwitz formula for the cover is:

$$2p - 2 = n(2q-2) + \text{ram} > 2nq - 2n + 4n - 4$$

or

$$p + 1 > n(q+1).$$

Consequently, $g \circ f$ is meromorphic on $X_p$; its only pole is at $z$ and its order is less than or equal to $n(q+1)$. Consequently its order is less than or equal to $p$.

q.e.d.

Proof of Lewittes' Theorem. Let $T$ be the automorphism of order $n$ with 5 or more fixed points. Since the ramification of the cover $X_p \to X_p/\langle T\rangle$ is $5(n-1)$ or more, the lemma shows every fixed point to be a Weierstrass point.

### 6.2. $G_{168}$ on $X_3$. (Some folklore.)

Let $X_3$ be a Riemann surface of genus 3 admitting the finite (2,3,7) group $G_{168}$ of automorphisms of order 168. Consider the Galois covering $W_3 \to \mathbb{P}^1$ of 168 sheets. Let $z_2 \in \mathbb{P}^1$ be the point over which there are 84 branch points of multiplicity 2; let $z_3 \in \mathbb{P}^1$ be the point over which there are 56 branch points of multiplicity 3; let $z_7 \in \mathbb{P}^1$ be the point over which there are 24 branch points of multiplicity 7.

We claim that the 24 points above $z_7$ are the 24 ordinary

Weierstrass points in $X_3$. For the weights of points in a single G-orbit must all be the same. But the total Weierstrass weight is 24. The only place the Weierstrass points can fit is over $z_7$, all other orbits having more than 24 points.

Consider the points over $z_3$. The stabilizer of one of these points is an automorphism of order 3; call it T. If the genus of $W_3/\langle T\rangle$ is zero, there would be 5 fixed points, by the Riemann-Hurwitz formula, and so all would be ordinary Weierstrass points, by Lewittes' theorem, a contradiction to the previous paragraph. Consequently, the genus of $X_3/\langle T\rangle$ $(= X_1)$ must be one, and the Galois covering of 3 sheets must have 2 fixed points. Let $2x + 2y$ be the divisor of branch points in the covering $X_3 \to X_1$. Then this is a canonical divisor on $X_3$, and corresponds to the intersection of a line with $\varphi(X_3)$ where $\varphi : X_3 \to \mathbb{P}^2$ is the map given by the canonical series onto a non-singular plane quartic $C_4$. Thus $2x + 2y$ corresponds to the two points of contact of a bitangent to $C_4$. Since there are 56 points above $z_3$, these points are the 56 points of contact of the 28 bitangent to $C_4$.

We now locate the $2^{nd}$ order Weierstrass points (which we call 2K points).

The total weight of the 2K points is 108. Let $w_2, w_3$, and $w_7$ be the 2K weight of a single point above $z_2, z_3$, and $z_7$ respectively. (Since all other G-orbits have order 168, none of these points can have positve 2K weight.) Thus

$$(*) \qquad 84w_2 + 56w_3 + 24w_7 = 108 .$$

The only possible solution in positive integers to this equation is

$$w_2 = 1, \; w_3 = 0, \text{ and } w_7 = 1 .$$

The total weight of the kK points is $(2k - 1)^2 \cdot 12$. Let $w_0$ be the kK weight above points of $\mathbb{P}^1 - \{z_2, z_3, z_7\}$. Thus if $w_2, w_3$, and $w_7$ are the kK weights above $z_2, z_3$, and $z_7$ we have $168w_0 + 84w_2 + 56w_3 + 24w_7 = (2k - 1)^2 \cdot 12$. If we reduce this equation module 7 we have:

$$3w_7 \equiv 5(2k - 1)^2 \pmod 7$$

or

$$w_7 \equiv 4(2k - 1)^2 \pmod 7 .$$

Since $0 \le w_7 \le 5$ we see that $w_7$ is the smallest non-negative residue of $4(2k-1)^2$ (mod7). If we let $w_7(k)$ $(k \ge 2)$ be the $kK$ weight of a point above $z_7$, we see that $w_7(2) = 1$, $w_7(3) = 2$, $w_7(4) = 0$, $w_7(5) = 2$, $w_7(6) = 1$, $w_7(7) = 4$, $w_7(8) = 4$, and $w_7(k)$ is periodic in $k$ with period 7 for $k \ge 2$.

6.3. Cyclic coverings of $\mathbb{P}^1$ of prime order: Examples.
(More folklore.) [3].

The title of this section refers to the Riemann surface for the polynomial $P(x,y) = y^n - \prod_{j=0}^{s} (x - a_j)^{f_j}$ where $n$ is prime, $0 < f_j < n$ and (for convenience) we assume $\sum_1^s f_j \equiv 0 \pmod{n}$.

Then the n-sheeted covering $\varphi : X_p \to \mathbb{P}^1$ for this polynomial has branching at points above the $a_j$'s of multiplicity $n$ and is smooth over $\infty$. Thus the Riemann-Hurwitz formula gives $2p = (s-2)(n-1)$. Let $A_j$ be the point on $W_p$ above $a_j$. Our problem is to discover to what extent the $A_j$'s are higher order Weierstrass points. In this section we shall work out some examples which will motivate the more general discussion that will follow.

On $X_p$, the Riemann surface for $P(x,y)$ above, there are functions $y$ and $\varphi^*(x)$, which we will call simply $x$, where $M(X_p) = \mathbb{C}(x,y)$. Also we have a meromorphic differential, the lift of $dx$ on $\mathbb{P}^1$, which we shall also denote by $dx$. We wish to find the divisors of these objects on $X_p$.

Since $\sum_{j=1}^{s} f_j \equiv 0 \pmod{n}$ the covering $\varphi : X_p \to \mathbb{P}^1$ is smooth over $\infty$. Call the $n$ points on $X_p$ over $\infty, \infty_1, \cdots, \infty_n$. Thus the function $x - x_0$ on $\mathbb{P}^1$ lifts to a function with $n$ zero above $x_0$ and simple poles at $\infty_1, \cdots, \infty_n$. $dx$ lifted has a zero of order $n-1$ at each $A_j$ and $n$ poles of order 2 at each point above $\infty$. Near $\varphi^{-1}(\infty)$, $y = \sqrt[n]{\prod (x-a_j)^{f_j}}$ so that at each $\infty_i$,

$y$ has a pole of order $\sum f_j/n$. Thus over $\infty$, $y$ has a total of

$\sum f_j$ poles. Near $A_j$, $y = (x-a_j)^{(f_j/n)}$ $k(x)$ where $k(a_j) \neq 0$.
A local parameter $t$ at $A_j$ can be choosen so that $t^n = x - a_j$; so we see that $y$ has a zero of order $f_j$ at $A_j$.

We can summarize this information about $x, y$ and $dx$ on $X_p$ by the following table:

| | $A_1$ | $A_2$ | ... | $A_s$ | total over $\infty$ | total over $\varphi^{-1}(x_0)$ |
|---|---|---|---|---|---|---|
| $dx$ | $n-1$ | $n-1$ | | $n-1$ | $-2n$ | |
| $y$ | $f_1$ | $f_2$ | ... | $f_s$ | $-\sum f_j$ | |
| $x-x_0$ | 0 | 0 | | 0 | $-n$ | $n$ |

As an example consider the polynomial

$$P(x,y) = y^5 - (x-a_1)(x-a_2)(x-a_3)(x-a_4)^2 \qquad (p = 4)$$

| $K = g^3_6$ | $A_1$ | $A_2$ | $A_3$ | $A_4$ | total over $\infty$ |
|---|---|---|---|---|---|
| $dx$ | 4 | 4 | 4 | 4 | -10 |
| $y$ | 1 | 1 | 1 | 2 | -5 |
| $x-a_1$ | 5 | 0 | 0 | 0 | -5 |
| $\omega_1: dx/y^2$ | 2 | 2 | 2 | 0 | 0 |
| $\omega_2: (x-a_4)\,dx/y^3$ | 1 | 1 | 1 | 3 | 0 |
| $\omega_3: (x-a_4)\,dx/y^4$ | 0 | 0 | 0 | 1 | 5 |
| $\omega_4: (x-a_4)^2\,dx/y^4$ | 0 | 0 | 0 | 6 | 0 |
| $\omega_5: (x-a_1)(x-a_4)\,dx/y^4$ | 5 | 0 | 0 | 1 | 0 |

The last 5 rows correspond to holomorphic one-forms on $X_4$, $\omega_1, \omega_2, \cdots, \omega_5$. We see that all $A_j$'s are ordinary Weierstrass points with $A_1, A_2$, and $A_3$ having weight 2 and $A_4$ having weight 4.

By multiplying together two holomorphic one-forms we can compute non-gaps for quadratic differentials as follows:

| $2K = g^{8}_{12}$ | $A_1$ | $A_2$ | $A_3$ | $A_4$ | total over $\infty$ |
|---|---|---|---|---|---|
| $\omega^2_1$ | 4 | 4 | 4 | 0 | 0 |
| $\omega^2_2$ | 2 | 2 | 2 | 6 | 0 |
| $\omega^2_3$ | 0 | 0 | 0 | 2 | 10 |
| $\omega^2_4$ | 0 | 0 | 0 | 12 | 0 |
| $\omega^2_5$ | 10 | 0 | 0 | 2 | 0 |
| $[(x-a_2)/(x-a_1)]\,\omega^2_5$ | 5 | 5 | 0 | 2 | 0 |
| $\omega_1\omega_2$ | 3 | 3 | 3 | 3 | 0 |
| $\omega_1\omega_5$ | 7 | 2 | 2 | 1 | 0 |
| $\omega_2\omega_5$ | 6 | 1 | 1 | 4 | 0 |
| $\omega_3\omega_4$ | 0 | 0 | 0 | 7 | 5 |
| $\omega_2\omega_4$ | 1 | 1 | 1 | 9 | 0 |

The 2K non-gaps at $A_1$ are $0,1,2,3,4,5,6,7,10$; weight 2.
The 2K non-gaps at $A_4$ are $0,1,2,3,4,6,7,9,12$; weight 8.

By multiplying one-forms with quadratic differentials we can obtain the 3K non-gaps at the $A_j$'s. And so forth.

The table on the following page for a slightly more complicated case is worked out. It is readily seen that the $kK$ weights at each $A_j$ are periodic with period 7 for $k \geq 2$. Several features of this table will later be shown to be special cases of more general phenomena.
A final useful observation is the following.

Let $y^n = \prod_{j=1}^{s} (x - a_j)^{f_j}$ define a Riemann surface whose function field is $\mathbb{C}(x,y)$. Then letting

$$z = y^{\delta} \prod_{j=1}^{s} (x - a_j)^{\alpha_j}, \quad \delta \not\equiv 0 \pmod{n}$$

we see that $\mathbb{C}(x,y) = \mathbb{C}(x,z)$ and

kK-nongaps for $y^7 = (x-a_1)(x-a_2)(x-a_3)^2(x-a_4)^3$ (p=6)

| | | nongaps | | | | | | | | | | | | weight |
|---|---|---|---|---|---|---|---|---|---|---|---|---|---|---|
| k = 1 | $a_1$ | | | 0 | 1 | 2 | 3 | 4 | | | 7 | | | 2 |
| | $a_3$ | | | 0 | 1 | 2 | 3 | | 5 | | | 8 | | 4 |
| | $a_4$ | | | 0 | 1 | 2 | | 4 | 5 | | | | 9 | 6 |
| k = 2 | $a_1$ | 0-8 | 9 | 10 | 11 | | 13 | 14 | 15 | | | | | 3 |
| | $a_3$ | 0-8 | 9 | 10 | 11 | | 13 | | | 16 | | 18 | | 8 |
| | $a_4$ | 0-8 | 9 | 10 | 11 | | 13 | 14 | | | | 18 | | 6 |
| k = 3 | $a_1$ | 0-18 | 19 | 20 | 21 | 22 | 23 | 24 | | | | | | 0 |
| | $a_3$ | 0-18 | 19 | 20 | 21 | | 23 | 24 | | 26 | | | | 4 |
| | $a_4$ | 0-18 | 19 | 20 | | 22 | 23 | | | 26 | 27 | | | 8 |
| k = 4 | $a_1$ | 0-28 | 29 | 30 | 31 | | 33 | | 35 | | 37 | | | 6 |
| | $a_3$ | 0-28 | 29 | | 31 | 32 | | 34 | | 36 | | | 39 | 12 |
| | $a_4$ | 0-28 | 29 | 30 | 31 | 32 | | | 35 | 36 | | | | 4 |
| k = 5 | $a_1$ | 0-38 | 39 | 40 | 41 | 42 | 43 | 44 | | | | | | 0 |
| | $a_3$ | 0-38 | 39 | 40 | 41 | 42 | | 44 | | | 47 | | | 4 |
| | $a_4$ | 0-38 | 39 | 40 | 41 | | | 44 | 45 | | | 48 | | 8 |
| k = 6 | $a_1$ | 0-48 | 49 | 50 | 51 | 52 | 53 | | | | 57 | | | 3 |
| | $a_3$ | 0-48 | 49 | 50 | | 52 | | 54 | 55 | | 57 | | | 8 |
| | $a_4$ | 0-48 | 49 | 50 | | 52 | 53 | 54 | | | 57 | | | 6 |
| k = 7 | $a_1$ | 0-58 | 59 | 60 | 61 | | 63 | 64 | | | | | 70 | 8 |
| | $a_3$ | 0-58 | 59 | 60 | | 62 | 63 | | 65 | | | | 70 | 10 |
| | $a_4$ | 0-58 | 59 | | 61 | 62 | 63 | | | 66 | | | 70 | 12 |
| k = 8 | $a_1$ | 0-68 | | 70 | 71 | 72 | 73 | 74 | | | 77 | | | 8 |
| | $a_3$ | 0-68 | | 70 | 71 | 72 | 73 | | 75 | | | 78 | | 10 |
| | $a_4$ | 0-68 | | 70 | 71 | 72 | | 74 | 75 | | | | 79 | 12 |

$$z^n = \prod_{j=1}^{s} (x - a_j)^{\delta f_j + n\alpha_j}$$

defines the same covering of $\mathbb{P}^1$. Thus

$$y^7 = (x - a_1)(x - a_2)(x - a_3)^5$$
$$y^7 = (x - a_1)^2(x - a_2)^2(x - a_3)^3$$
$$y^7 = (x - a_1)^3(x - a_2)^3(x - a_3)$$
$$y^7 = (x - a_1)^4(x - a_2)^4(x - a_3)^2 \quad \text{etc.}$$

all define the same covering $x : X_3 \to \mathbb{P}^1$.

6.4. Lemma. Let $y^n = \prod_{j=1}^{s} (x - a_j)^{f_j}$ where $n$ is prime, $0 < f_j < n$, and $\sum f_j \equiv 0 \pmod{n}$. Let $A_j$ be the point on the Riemann surface $X_p$ above $a_j$. Then $n(2p - 2)A_j \equiv nK$ for all $j$ and so $w(nK, A_j) > 0$ for all $j$.

Proof. Let $g^1_n$ be the fibers of the cover $X_p \to \mathbb{P}^1$. Then by Theorem 2.6 we have ($B$ is the branched locus)

$$K \equiv -2g^1_n + B$$

where we write the result additively. Now $g^1_n \equiv nA_1$, and

$$B = \sum_{j=1}^{s} (n - 1)A_j .$$

$$\text{Thus } nK \equiv -2ng^1_n + \sum_{j=1}^{s} (n - 1)\, nA_j \equiv -2n^2 A_1 + s(n - 1)\, nA_1$$
$$= n\big(-2n + s(n - 1)\big)A_1 = n(2p - 2)A_1 .$$

q.e.d.

This lemma motivates most of the work in this chapter.

6.5. We first review some material form Chapter 2, with slightly different notation. Let $|G| = g^R_N$ be a complete linear series without base point where $G$ is an integral divisor on $X_p$. For $z \in X_p$ there are $R+1$ G-non-gaps for $g^R_N$ at $z$, $0 = n_0 < n_1 < \cdots < n_R \leq N$. The $N - R$ $(= T)$ G-gaps at $z$ are $g_1 < g_2 < \cdots < g_T \leq N$.

$T = N - R = p - i_0$ where $i_0$ is the index of speciality of $g^R_N$. For all $j = 0,1,\cdots,R$

$$G \equiv n_j z + E_j$$

where $|E_j| = g^{R-j}_{N-n_j}$ and $(E_j, z) = 0$. Then the generalized Weierstrass weight of $|G|$ at $z$, denoted $w(G,z)$, and called simply the <u>G-weight at z</u> is:

$$w(G,z) = \sum_{j=0}^{R} (n_j - j) = \sum_{j=1}^{T} (R + j - g_j) = \sum \hat{n} - \hat{g}$$

where $\hat{n}$ stands for the extraordinary non-gaps $n_j > R$ and $\hat{g}$ stands for the extraordinary gaps $g_j \leq R$.

<u>6.6</u>. <u>Lemma</u>. The following three conditions are equivalent.

(1) $g$ is a G-gap at $z$.

(2) $i(G - gz) + 1 = i\big(G - (g+1)z\big)$.

(3) There exists an integral divisor $E$ of degree $g + 2p - N - 1$ so that $\big(|E|, z\big) = 0$ and

$$G + E \equiv (g+1)z + K.$$

(Remark: It is not required that $G - gz$ be integral.)

<u>Proof</u>. (1) $\Leftrightarrow$ (2) : $g$ is a G-gap at $z \Leftrightarrow r(G - gz) = r\big(G - (g+1)z\big)$ $\Leftrightarrow N - g - p + i(G - gz) = N - (g+1) - p + i\big(G - (g+1)z\big)$.

(2) $\Leftrightarrow$ (3) : $i\big(G - (g+1)z\big) > 0$. Therefore there exists an integral divisor $E$ so that $G - (g+1)z + E \equiv K$ or $G + E \equiv (g+1)z + K$. Now $z$ is a fixed point of $|E| \Leftrightarrow r(E - z) = r(E) \Leftrightarrow i(G - gz) = i\big(G - (g+1)z\big)$. Therefore $z$ is not a fixed point of $|E|$. Now reverse the argument.

q.e.d.

(Note: If $G = K$ we get the usual results about ordinary Weierstrass points.)

<u>6.7</u>. <u>Lemma</u>. Let $i_0 = i(G)$. Then $i_0 = i(G - g_1 z)$ and $i\big(G - (g_j + 1)z\big) = i_0 + j$, $j = 1,2,\cdots,T$.

<u>Proof</u>. Because $G_1$ is the first gap we have $r(G - g_1 z) = R - g_1$.

Therefore $|G - g_1 z| = g^{R-g_1}_{N-g_1}$, and it is complete; thus it has the same index as $|G|$.

By the previous lemma, if $n$ is a non-gap then $i(G - nz) = i\big(G - (n+1)z\big)$. Thus the index of $G - nz$ changes only at the gaps, and at each gap the change is one.

q.e.d.

6.8. Lemma. If $|G|$ is not special then for $m$ a positive integer $w(G + mz, z) = w(G,z)$.

Proof. It suffices to prove the lemma for $m = 1$ (and then use induction).

For any j $\quad G \equiv n_j z + E_j \quad (E_j, z) = 0$

$$G + z \equiv (n_j + 1)z + E_j .$$

Now $|G + z| = g^{R+1}_{N+1}$, so the $(G + z)$-non-gaps are

$$0, n_0 + 1, n_1 + 1, \cdots, n_R + 1$$

and the $(G + z)$-gaps are

$$g_1 + 1, \cdots, g_T + 1 .$$

Thus the sum $\sum \hat{n} - \hat{g}$ is unchanged.

6.9. Lemma. The smallest G-gap at $z, g_1$, satisfies:

$$g_1 \geq N - 2p + 1 .$$

Equality occurs $\Leftrightarrow G \equiv (g_1 + 1)z + K$

(in which case $g_1 + 1$ is a non-gap).

Proof. $i_0 = i(G - g_1 z)$. If $i_0 > 0$ then $N \leq 2p - 2$ and there is nothing to prove. Therefore suppose $i_0 = 0$. Then $i\big(G - (g_1 + 1)z\big) = 1$ so $\deg\big(G - (g_1 + 1)z\big) \leq 2p - 2$ or $N - (g_1 + 1) \leq 2p - 2$.

Equality occurs $\Leftrightarrow G - (g_1 + 1)z \equiv K$ since $K$ is characterized by having degree $2p - 2$ and index $1$.

6.10. Lemma. If $m \geq 2$ then $w(K + mz, z) = w(K,z) + p$.

Proof. Let $G = mz + K$. Then $|G| = g^R_N$ where $N = m + 2p - 2$, $N - R = p$. By Lemma 6.9 $g_1 = m - 1$. Let $n'_0, \cdots, n'_{p-1}$ be the K-non-gaps at $z$. Since $K$ is fixed-point-free there are $p$ G-non-gaps

$$m + n'_j \qquad j = 0,1,\cdots,p-1 .$$

Thus the $m + p - 1$ G-non-gaps are

$$0,1,\cdots,m-2\ ,\ m+n'_0,\cdots,m+n'_{p-1} .$$

$$\text{Thus } w(G,z) = \sum_{j=0}^{p-1} (m + n'_j) - (m - 1 + j)$$

$$= p + \sum_{j=0}^{p-1} (n'_j - j) = p + w(K,z) .$$

6.11. Lemma. Suppose $G = Nz$ where $|G| = g^R_N$, $N - R = p$. Then (1) $g_p = N - 1$ and (2) $w(G,z) = p + w(K,z)$.

Proof. (1) Since $i_0 = 0$ Lemma 6.7 gives, for $j = p$ $i\big((N - g_p - 1)z\big) = p$. Since $K$ is fixed-point-free this implies $N - g_p - 1 = 0$.

(2) By Lemma 6.7 $i\big((N - g_j - 1)z\big) = j$.

Thus $\{N - g_j - 1 \mid j = 1,2,\cdots,p\}$ are the K-non-gaps at $z$.

$$w(K,z) = \sum_0^{p-1} n'_j - j = \sum_1^{p} (N - g_j - 1) - (j - 1)$$

$$= \sum (N - p + j) - g_j - 1 \quad \Big(\sum_1^p j = \sum_1^p p + 1 - j\Big)$$

$$= w(G,z) - p . \qquad (N - p = R)$$

6.12. Lemma. Let $G = Nz$, $\deg H = M$, $M \ge 2$, and $|G - H|$ is not special. Then $w(G - H,z) = w(H + K,z) = w(G + H + K,z)$.

Proof. Since $G = Nz$ the second equality follows form Lemma 6.8. For the first equality, let $g_j$ be a $(G - H)$-gap. By Lemma 6.6 there exists an integral divisor $E_j$, $(E_j,z) = 0$ and

$$G - H + E_j \equiv (g_j + 1)z + K$$

or

$$(N - g_j - 1)z + E_j \equiv H + K .$$

Thus $N - g_j + 1$ is an $(H + K)$-non-gap. Now $g_j \le N - M$

so $$M-1 \le N-g_j-1 .$$

$|H+K| = g^{M+p-2}_{M+2p-2}$. There are $M+p-1$ $(H+K)$-non-gaps of which the first $M-1$ are $0,1,\cdots,M-2$ (Lemma 6.9). The other $(H+K)$-non-gaps are $N-g_j-1$, $j=1,2,\cdots,p$. Compute $w(H+K,z)$ for non-gaps $\ge M-1$.

$$w(H+K,z) = \sum_{j=1}^{p}\Big[(N-g_j-1)-\big((M-1+(j-1)\big)\Big]$$

$$= \sum_{j=1}^{p}\big(N-M-(j-1)-g_j\big) .$$

But $$w(G-H,z) = \sum_{j=1}^{p}(N-M-p+j-g_j) .$$

q.e.d.

<u>6.13</u>. In the light of Lemma 6.4 the following theorem explains many features of the results tabulated in Section 6.3.

<u>Theorem 6.13</u>. Suppose for $z \in X_p$ we have $m(2p-2)z \equiv mK$.

(1) If $m=1$, then for all $\ell = 2,3,4,\cdots$ we have
$$w(\ell K,z) = w(K,z)+p .$$
(2) If $k,\ell \ge 2$ and $k \equiv \ell \pmod m$ then
$$w(kK,z) = w(\ell K,z) .$$
(3) If $m \ge 2$, $k+\ell \equiv 1 \pmod m$, and $k,\ell \ge 2$ then
$$w(kK,z) = w(\ell K,z) .$$
(4) If $m \ge 2$, $\ell = 1,2,\cdots$ then
$$w(\ell mK,z) = w\big((\ell m+1)K,z\big) = w(K,z)+p .$$

<u>Proof</u>. (1) follows from Lemma 6.11.

(2) follows from Lemma 6.8.

(3) follows from Lemma 6.12. For any $t \ge 1$ and $k$ so that $2 \le k \le tm-2$ let $G = tmK$ and $H = kK$ in Lemma 6.12.

(4) follows from Lemma 6.11 and (3) above.

<u>6.14</u>. Cyclic converings in general.

We now show how to generalize the results of Theorem 6.13 to cyclic coverings of prime order where the quotient is no longer required to be the Riemann sphere. Thus we consider a Riemann surface $X_p$ of genus $p$ with an automorphism $T$ of prime order $n$ and $X_q = X_p/\langle T\rangle$. The branched covering $\varphi : X_p \to X_q$ will have $n$-sheets. If $M(X_q)$ is the field of meromorphic functions on $X_q$ then we will denote $\varphi^*\big(M(X_q)\big)$ by $M^*(X_q)$ a subfield of $M(X_q)$ of index $n$.

Lemma 6.14. There exists $y \in M(X_p) - M^*(X_q)$, with $y^n \in M^*(X_q)$ and $M(X_p) = M^*(X_q)[y]$.

Proof. Pick $y \in M(X_p) - M^*(X_q)$. Since $n$ is prime,

$M(X_p) = M^*(X_q)[y]$. Let $\tau = e^{2\pi i/n}$. Then $\sum_{m=0}^{n-1} \tau^{km} = 0$ if

$k = 1,\cdots,n-1$. For $k = 0,1,\cdots,n-1$ let $y_k = \sum_{m=0}^{n-1} \tau^{-km}(y \circ T^m)$.

Thus $y_k \circ T = \tau^k \sum_{m=0}^{n-1} \tau^{-k(m+1)}(y \circ T^{m+1}) = \tau^k y_k$.

Also $\sum_{k=0}^{n-1} y_k = \sum_{m=0}^{n-1}\Big(\sum_{k=0}^{n-1} \tau^{-km}\Big)\, y \circ T^m = ny$.

Therefore, there is a $y_k$, not in $M^*(X_q)$, so that $y_k^n \in M^*(X_q)$ since $(y_k \circ T)^n = y_k^n \circ T = y_k$.

q.e.d.

We can and will assume that $y$ is chosen as in this lemma so that $y \circ T = \tau y$. Let $y^n = F \in M^*(X_q)$. Then $F = \tilde{F} \circ \varphi$ for a function $\tilde{F} \in M(X_q)$. The divisor of $\tilde{F}$ can be written

$(\tilde{F}) = \sum_{j=1}^{s} f_j a_j + nE_0$ where $E_0$ is a divisor on $W_q$ and

$0 < f_j < n-1$. Because $\deg(\tilde{F}) = 0$, we have $\sum f_j \equiv 0 \pmod{n}$. (Some of the $a_j$'s can also occur in $E_0$.) Since $X_p$ is the Riemann surface for the polynomial $X^n - \tilde{F} \in M(X_q)[X]$ we see that the covering $\varphi : X_p \to X_q$ is branched over each $a_j$ and only there. (If $(\tilde{F}) = nE_0$, then the covering is smooth.)

The ramification over each $a_j$ is $n-1$ so the Riemann-Hurwitz formula for the covering is

$$2p-2 = n(2q-2) + s(n-1) .$$

Let $A_j$ be the point on $X_p$ above $a_j$ on $X_q$. Then the divisor of $y$ is

$$(y) = \sum_{j=1}^{s} f_j A_j + \varphi^{-1}(E_0)$$

and $\sum f_j \equiv 0 \pmod{n}$.

<u>6.15</u>. <u>Lemma</u>. Suppose $g \in M(X_p)$ where $\varphi : X_p \to X_q$ is an n-sheeted covering of compact Riemann surfaces. Suppose there is a divisor of degree zero on $X_q$, $D_0$, so that $(g) = \varphi^{-1}(D_0)$. Then $g^n \in \varphi^* M(X_q)$. (We do not assume that n is prime or that the covering is Galois.)

<u>Proof</u>. For $y \in X_q$, let $\varphi^{-1}(y) = \{x_1, \cdots, x_n\}$. Define

$$G(y) = \prod_{j=1}^{n} g(x_j) \text{ , a well defined function on } X_q \text{ . Then } (G) = n D_0 .$$

So $G \circ \varphi$ and $g^n$ have the same divisor.

q.e.d.

<u>6.16</u>. Invariant divisors.

<u>Definition</u>. Let H be a group of automorphisms on a Riemann surface X. A linear series $g^R_N$ will be said to be <u>invariant under H</u> if for all $h \in H$ and all divisors D in $g^R_N$, we have $hD \in g^R_N$.

For example the k-canonical series will be invariant under any subgroup of $\mathrm{Aut}(X)$.

<u>Lemma 6.16</u>. Let $g^R_N$ be a complete linear series invariant under a cyclic group $\langle T \rangle$ of prime order. Then there exists a divisor $G_0$ in $g^R_N$ so that $TG_0 = G_0$.

The proof of this lemma will depend on the following general fact.

<u>Proposition 6.16</u>. Let $|G| = g^R_N$ be a complete linear series invariant under a group $H \in \mathrm{Aut}(X)$. Consider a map

$$\theta : X \to \mathbb{P}^R$$

given by $R+1$ independent functions in $L(G)$. Then there is a group of projective transformations in $\mathbb{P}^R$, $\hat{H}$, isomorphic to $H$ and $\theta(X)$ is invariant under $\hat{H}$. Moreover, $\hat{H}$ acts on the hyperplanes of $\mathbb{P}^R$ in the same manner that $H$ acts on the divisors in $|G|$.

Proof. Let $\langle f_0, \cdots, f_R \rangle$ be a basis of $L(G) = \{f \in M(X) \mid (f) + G > 0\}$. Then $\langle f_0 \circ T, \cdots, f_R \circ T \rangle$ is a basis of $L(TG)$ where $T \in H$. Since $g^R_N$ is invariant under $T$, there exists $F \in M(X)$ and

$$(F) = G - TG.$$

Thus $(f_k \circ T)F \in L(G)$ and so

$$(f_k \circ T)F = \sum_{j=0}^{R} a_{k_j} f_j \qquad (a_{k_j})^{(R+1)\times(R+1)}.$$

Define $\theta : X \to \mathbb{P}^R$ by $x \to (f_0(x), \cdots, f_R(x))$. Then $Tx$ maps into

$$\begin{aligned} &(f_0 \circ T(x), \cdots, f_R \circ T(x)) \\ &= ((f_0 \circ T(x))F(x), \cdots, (f_R \circ T(x))F(x)) \\ &= \left(\sum_{j=0}^{R} a_{0j} f_j, \cdots, \sum_{j=0}^{R} a_{Rj} f_j\right). \end{aligned}$$

Thus $T$ induces on $\mathbb{P}^R$ a projective transformation $T$ given with respect to this basis by the matrix $(a_{kj})$. Everything now follows.

q.e.d.

Proof of Lemma 6.16. Suppose $g^R_N$ is invariant under $\langle T \rangle$. Then $\hat{T}$ acting on $\mathbb{P}^R$ also acts on $\mathbb{P}^{R*}$, the hyperplanes in $\mathbb{P}^R$. Suppose the action of $\hat{T}$ on $\mathbb{P}^{R*}$ is given by the matrix $\widetilde{M}$. Thus $\widetilde{M}^n = \lambda I$, $\lambda \in \mathbb{C}$. If $M = \frac{1}{\sqrt[n]{\lambda}} \widetilde{M}$ then $M^n = I$; that is, we can lift the cyclic group to $\mathbb{C}^{R+1} - \{0\}$ over $\mathbb{P}^R$. Now we proceed as before. For $v \in \mathbb{C}^{R+1}$ define

$$v_k = \sum_{m=0}^{n-1} \tau^{-mk} M^m v$$

and so $Mv_k = \tau^k v_k$ and $nv = \sum v_k$ .

Therefore there is a $v_k \neq 0$ . The line through the origin and $v_k$ in $\mathbb{C}^{R+1}$ corresponds to a point in $\mathbb{P}^{R*}$ , which in turn corresponds to an invariant hyperplane $H$ in $\mathbb{P}^R$ . The pull back to $X$ of the hyperplane section $\theta(X) \cap H$ will be a divisor in $|G|$ invarant under $T$ (but not necessarily pointwise invariant).

q.e.d.

6.17. We continue the previous notation, some of which we now recapitulate. $T$ is an automorphism of $X_p$ of prime order $n$ , and $\varphi : X_p \to X_q$ is the quotient map. $y$ generates $M(X_p)$ over $M^*(X_q)$ where $y^n \in M^*(X_q)$

$$(y) = \sum_{j=1}^{s} f_j A_j + \varphi^{-1}(E_0)$$

where $E_0$ is a divisor on $X_q$ . We will not assume $s > 0$ , although most of the complications which follow are the result of assuming $T$ has fixed points.

Let $g^R_N$ be a complete linear series invariant under $\langle T \rangle$ . We will assume for convenience that $g^R_N$ is without base points; although this assumption is not necessary . Let $G_0$ be an invariant divisor in $G^R_N$ ; that is $TG_0 = G_0$ . Then

$$G_0 = \sum_{j=1}^{s} g_j A_j + \varphi^{-1}(D_0)$$

where $0 \leq g_j < n$ and $N \equiv \sum_{j=1}^{s} g_j \pmod{n}$ .

For example if $g^R_N = K_p$ , the canonical series on $X_p$ , then

$$G_0 = \sum_{j=1}^{s} (n-1) A_j + \varphi^{-1}(K_q) .$$

If $g^R_N = |kK_p|$ then for $1 \le k \le n$

$$G_0 = \sum_{j=1}^{s} (n-k)A_j + \varphi^{-1}\Big(kK_q + (k-1)(a_1 + \cdots + a_s)\Big).$$

$a_j = \varphi(A_j)$.

Let $L = L(G_0) = \left\{F \in M(X_p) \,\middle|\, (F) + G_0 \ge 0\right\}$. Then $L(G_0)$ is invariant under $T$ and $\dim L = R + 1$. Let $L_m = \left\{F \in L \,\middle|\, F \circ T = \tau^m F\right\}$, $m = 0, 1, \cdots, n$. Let $d_m = \dim L_m$.

Then $L = \sum_{m=0}^{n-1} L_m$ and $R + 1 = \sum_{m=0}^{n-1} d_m$.

Let $\mathcal{S}_m = \left\{(F) + G_0 \,\middle|\, F \in L_m\right\}$ where $\mathcal{S}_m = \emptyset$ if $d_m = 0$.

We want a more explicit description of the divisors in $\mathcal{S}_m$.

Note that $(y^m) = \sum_{j=1}^{s} mf_jA_j + \varphi^{-1}(mE_0)$.

If $F \in L_m$ then $(F/y^m) \circ T = F/y^m$; that is $F/y^m \in M^*(X_q)$. Suppose $(F) = G_m - G_0$ where $G_m \in \mathcal{S}_m$.

$$(F/y^m) = \varphi^{-1}(D) \text{ for some divisor } D \text{ in } X_q$$

thus $G_m - G_0 = (y^m) + \varphi^{-1}(D)$

$$G_m = G_0 + (y^m) + \varphi^{-1}(D).$$

$$G_m = \sum g_jA_j + \varphi^{-1}(D_0) + \sum mf_jA_j + \varphi^{-1}(mE_0) + \varphi^{-1}(D)$$

$$= \sum (mf_j + g_j)A_j + \varphi^{-1}(D')$$

$$= \sum t_{mj}A_j + \varphi^{-1}(D'')$$

where $t_{mj} = \overline{mf_j + g_j}$, the smallest non-negative residue of $mf_j + g_j \pmod n$, $D'$ and $D''$ are divisors on $X_q$, and $D'' > 0$.

<u>Lemma 6.17</u>. If $\mathcal{S}_m$ is non-empty then $\mathcal{S}_m$ is the set of all divisors

of the form $\sum t_{mj} A_j + \varphi^{-1}(D_m)$ where a) the divisor is linearly equivalent to $G_0$, and b) $D_m$ is integral.

Proof. That any divisor with these proeprties is in $\mathcal{S}_m$ follows by reversing the above argument.

Theorem 6.17. For each $m = 0,1,\cdots,n-1$ there exists on $X_q$ a divisor $D_m$, not nessarily integral, so that

a) $G_0 \equiv \sum_{j=1}^{s} t_{mj} A_j + \varphi^{-1}(D_m)$ and

b) $d_m = \dim |D_m|$.

Proof. For part a) we unwind the previous calculation.

$$G_0 - \sum_{j=1}^{s} t_{mj} A_j = \sum g_j A_j + \varphi^{-1}(D_0) - \sum (g_j + mf_j)A_j - \varphi^{-1}(B)$$

$$= - \sum_{j=1}^{s} mf_j A_j - m\varphi^{-1}(E_0) + \varphi^{-1}(D_m)$$

$$= -(y^m) + \varphi^{-1}(D_m)$$

where $D_m = mE_0 + D_0 - B$ and $B$ is a divisor supported in $\{a_1,\cdots,a_s\}$. Thus

$$G_0 \equiv \sum_j t_{mj} A_j + \varphi^{-1}(D_m).$$

Now $\dim |D_m| = 0$ if and only if there is no integral divisor equivalent to $D_m$. By the lemma this is equivalent to $\mathcal{S}_m = \emptyset$, which in turn is equivalent to $d_m = 0$.

If $\dim |D_m| \geq 1$, let $D_m^0$ be a fixed divisor in $|D_m|$. Then fix $F^0 \in L_m$ where

$$(F^0) = \sum t_{mj} A_j + \varphi^{-1}(D_m^0) - G_0.$$

We define an isomorphism of $L(D_m^0)$ onto $L_m$ as follows.

If $(f) + D^0_m > 0$ let $D'_m = (f) + D^0_m$. Then

$$(F^0\varphi^* f) = \sum_{j=1}^{s} t_{mj} A_j + \varphi^{-1}(D'_m) - G_0 ;$$

that is, $F^0\varphi^* f$ is in $L_m$. The map $f \to F^0\varphi^* f$ is the required isomorphism. For if $F \in L_m$ then $F/F^0$ is seen to be $\varphi^* f$ for some $f \in L(D^0_m)$.

q.e.d.

6.18. Lemma. Let $r = s(n-1)$, the ramification of the covering $\varphi : X_p \to X_q$. Then $\sum_{m=0}^{n-1} \sum_{j=1}^{s} t_{mj} = n\frac{r}{2}$.

Proof. As $m$ runs through the integers $0,1,\cdots,n-1$ $t_{mj} = \overline{g_j + mf_j}$ does also since $f_j \not\equiv 0 \pmod{n}$. Thus

$$\sum_{j=1}^{s} \sum_{m=0}^{n-1} t_{mj} = \sum_{j=1}^{s} \frac{n(n-1)}{2} = n\frac{s(n-1)}{2} = n\frac{r}{2} .$$

q.e.d.

In Theorem 6.17 $\deg D_m = \left(N - \sum_{j=1}^{s} t_{mj}\right)/n$. Applying the Riemann-Roch theorem to $|D_m|$ on $X_q$, we have

$$\dim|D_m| = \deg D_m - q + 1 + i_m$$

where $i_m$ is the index of $|D_m|$. Summing this over $m$ gives

$$R + 1 = \sum_m d_m = N - \left(\sum_m \sum_j t_{mj}\right)/n - n(q-1) + \sum_m i_m$$

$$= N - \frac{r}{2} - n(q-1) + \sum_m i_m$$

$$= N - p + 1 + \sum_{m=0}^{n-1} i_m$$

by the Riemann-Hurwitz formula. Another application of the Riemann-Roch theorem on $X_p$ gives this

Theorem 6.18. $i(g^R_N) = \sum_{m=0}^{n-1} i_m$

Corollary 6.18. If $g^R_N$ is not special then

$$d_m = \left(N - \sum_{j=1}^{s} t_{mj}\right)/n - q + 1$$

a non-negative number.

6.19. We now assume that $g^R_N$ is non-special and we fix $A_\ell$. Assuming $\mathcal{S}_m \neq \emptyset$ we ask how the divisors in $\mathcal{S}_m$ contribute to the $g^R_N$ non-gaps at $A_\ell$. Fix a $G_m$ in $\mathcal{S}_m$ where

$$G_m = \sum_j t_{mj} A_j + \varphi^{-1}(D_m)$$

and $|D_m| = g^{d_m-1}_{d_m-1+q}$. If $D_m = n'a_\ell + D'$, $D' \geq 0$, $(a_\ell, D') = 0$, then $t_{m\ell} + nn'$ is a $g^R_N$ non-gap at $A_\ell$. Thus the divisors in $\mathcal{S}_m$ contribute $d_m$ non-gaps at $A_\ell$ and all of these $d_m$ non-gaps are congruent to $t_{m\ell}$ (mod $n$). Since there are

$R + 1 = \sum d_m$ non-gaps at $A_\ell$ we see that the only non-gaps at $A_\ell$ congruent to $t_{m\ell}$ come from divisors in $\mathcal{S}_m$ If $a_\ell$ is not a $D_m$-point (to be expected in general) then the non-gaps congruent to $t_{m\ell}$ at $A_\ell$ are

$$t_{m\ell}, t_{m\ell} + n, t_{m\ell} + 2n, \cdots, t_{m\ell} + (d_m - 1)n.$$

Definition. Let $d_{m\ell}$ be the number of integers in $0, 1, \cdots, R$ congruent to $t_{m\ell}$ (mod $n$).

(We remark that $\sum_{m=0}^{n-1} d_{m\ell} = R + 1 = \sum_{m=0}^{n-1} d_m$.)

Writing $w(G,A_\ell)$ as $\sum n_j - j$ we obtain:

$$w(G,A_\ell) = \sum_{j=0}^{R} n_j - \sum_{j=0}^{R} j$$

$$= \sum_{m=0}^{n-1} \sum_{j=0}^{d_m-1} (t_{m\ell} + jn) - \sum_{m=0}^{n-1} \sum_{j=0}^{d_{m\ell}-1} (t_{m\ell} + jn)$$

$$= \sum_{m=0}^{n-1} \sum_{j=d_{m\ell}}^{d_m-1} (t_{m\ell} + jn)$$

$$= \sum_{m=0}^{n-1} \tfrac{1}{2}\Big(t_{m\ell} + d_{m\ell}n + t_{m\ell} + (d_m - 1)n\Big)\Big(d_m - d_{m\ell}\Big) .$$

Since $\sum_m (d_m - d_{m\ell}) = 0$ we obtain the following.

Theorem 6.19. If $w(D_m, a_\ell) = 0$ for all $m$ then

$$w(G,A_\ell) = \tfrac{1}{2} \sum_{m=0}^{n-1} \left[2t_{m\ell} + n(d_m + d_{m\ell})\right](d_m - d_{m\ell}) .$$

Definition. The right-hand side of the equation in this theorem will be called the expected weight and denoted $w_e(G,A_\ell)$.

Corollary 6.19. (1) $w(G,A_\ell) \geq w_e(G,A_\ell)$

(2) $w(G,A_\ell) \equiv w_e(G,A_\ell) \pmod{n}$.

Proof. If $a_\ell$ is a $D_m$-point for some $m$, the contribution of $\delta_m$ to $w(G,A_\ell)$ will increase by a multiple of $n$ over that computed in the theorem.

q.e.d.

6.20. Lemma. The formula for $w_e(G,A_\ell)$ in Theorem 6.19 depends only on $\ell, f_1, \cdots, f_s, g_1 \cdots, g_s$ provided $g^R_N$ is non-special.

Proof. We show $w_e(G,A_\ell)$ does not depend on $N$ or $q$.

Suppose first that $q$ is fixed and $g^{R'}_{N'}$ has the same $g_1,\cdots,g_s$ as $g^{R}_{N}$. Now $N \equiv \sum g \pmod{n}$ so $N \equiv N' \pmod{n}$. If $N' = N + n$ then $g^{R'}_{N'} = g^{R+n}_{N+n}$. Since

$$d_m = \left(N - \sum_{j=1}^{s} t_{mj}\right)/n - q + 1,$$

$d_m$ is increased by one. $d_{mj}$, the number of integers congruent to $t_{m\ell}$ in $[0,R]$ is also increased by one. Since $\sum_m (d_m - d_{m\ell}) = 0$ the result follows.

Now suppose $q$ is increased by one while $N$ is fixed. Then $d_m$ is decreased by one. $p$ is increased by $n$, $R$ is decreased by $n$ and so $d_{m\ell}$ is decreased by one.

q.e.d.

We now consider the extent to which fixed points of automorphisms are higher order Weierstrass points, that is, $kK$-points for $k \geq 2$.

Theorem 6.20. Suppose $p \geq 2$ and $T$ is an automorphism of prime order $n$ with three or more fixed points. Then each fixed point is an $\ell K$-point for $\ell \geq 2$ and $\ell \equiv 0 \pmod{n}$ or $\ell \equiv 1 \pmod{n}$. (The case $\ell \equiv 1 \pmod{n}$ appears in Farkas-Kra [11]).

Proof. Consider the cover $X_p \rightarrow X_q$ $(= X_p/\langle T\rangle)$.
By the above lemma we may apply the results of the case $X_{p-nq} \rightarrow \mathbb{P}^1$ provided $p - nq \geq 2$. (Since $s \geq 3$, $p - nq \geq 1$.) But then the results of Theorem 6.13 (4) apply.

The two cases when $p - nq = 1$ are (i) $n = 2$, $s = 4$ and (ii) $n = 3$ and $s = 3$. Then the two cases to consider are (i) $W_3 \rightarrow W_1$ where $T^2 = 1d$ and (ii) $W_4 \rightarrow W_1$ where $T^3 = 1d$. Here we use the formula in Theorem 6.19 directly. We will only work out the second case $W_4 \rightarrow W_1$.

In this case $f_1 = f_2 = f_3 = 1$. For $3K_4$ $(= g^{14}_{18})$ we may take $g_1 = g_2 = g_3 = 0$. Then $d_0 = 6$, $d_1 = 5$, and $d_2 = 4$ while $d_{01} = d_{11} = d_{21} = 5$. Then

$$w(3K_4, A_1) = \frac{1}{2}\Big(2\cdot 0 + 3(6+5)\Big)(6-5) + \frac{1}{2}\Big(2\cdot 1 + 3(5+5)\Big)(5-5)$$
$$+ \frac{1}{2}\Big(2\cdot 2 + 3(4+5)\Big)(4-5)$$
$$= 1\,.$$

Actually, as soon as we have $d_0 > d_{01}$ we know there is an extraordinary $3K_4$-non-gap at $A_1$ and so $A_1$ is a $3K_4$-point.

Similarly, $w_e(4K_4, A_1) = 1$.

q.e.d.

## Chapter 7. Automorphisms of Riemann surfaces, II; N(p).

7.1 In section 5.5 through 5.8 we discussed automorphism groups of order $84(p-1)$ on $X_p$. If $X_p$ admits a group of automorphisms of order $84(p-1)$ then we immediately know that $N(p) = 84(p-1)$. If, however, $N(p) < 84(p-1)$ then determining $N(p)$ involves two problems: the existence problem, to show there is a Riemann surface of genus p admitting a group of the conjectured order; and the elimination problem, to show that no Riemann surface of genus p admits a group of order greater than that conjectured.

The problem of determining the function $N(p)$ for all p appears to be extremely difficult. Here we shall show how to find some infinite sequences of genera, $p_n$, where $N(p_n)$ can be determined.

7.2 In showing the existence of a group of automorphisms whose order competes for $N(p)$, very often we will be extending an abelian group by a group of genus zero. In using Macbeath's method of lifting automorphisms, Theorem 4.13 simplifies matters considerably.

Theorem 7.2. For $p \geq 2$, $N(p) \geq 8(p+1)$.

Proof. Fix an integer $p \geq 2$. Let $\tau = e^{2\pi i/(2p+2)}$. Let $B = \{\tau^k \mid k = 0,1,\ldots,2p+1\}$, the vertices of a regular $(2p+2)$-gon inscribed in the unit circle. Let $D_{2p+2}$ be the dihedral group in $\mathbb{P}^1$ generatd by $Vz = z^{-1}$ and $Rz = \tau z$. $D_{2p+2}$ permutes the set B. If $\gamma_0$ is a path circling $\tau^0 = 1$, let $\gamma_k = R\gamma_0$, a little path circling $\gamma^k$. Let $Z_2 = \langle\varphi\rangle$. Let $\mu : \pi_1(\mathbb{P}^1 - B, 2) \to Z_2$ take each $\gamma_k$ onto $\varphi$. The kernel of $\mu$ in $H_1(\mathbb{P}^1 - B, Z)$ is

$$\{\sum_{j=0}^{2p+1} c_j\gamma_j \mid \sum c_j \equiv 0 \pmod 2\}$$

a subgroup of index 2 which is invariant under $D_{2p+2}$. The corresponding 2-sheeted covering, with points filled in above the $2p+2$ points of B, is a hyperelliptic Riemann surface $X_p$. $D_{2p+2}$ lifts to $X_p$ which together with the hyperelliptic involution yields a group of automorphisms of order $8(p+1)$.

q.e.d.

7.3. We continue in the same vein.

Theorem 7.3. If 3 divides p then $N(p) \geq 8(p+3)$.

Proof. Consider the symmetric group $S_4$ acting as the orientation preserving symmetries of the cube inscribed in the Riemann sphere. Lable the 8 vertices $a_1, a_2, \dots, a_8$ so that any two of the vertices with odd indices lie at opposite ends of a diagonal of one of the square faces of the cube. Let $A_1 = \{a_1, a_3, a_5, a_7\}$ and $A_2 = \{a_2, a_4, a_6, a_8\}$. The even indexed vertices have the same property. Let $A = A_1 \cup A_2$. For $z_0 \notin A$ and any positive integer $m$ define $\mu : \pi_1(\mathbb{P}^1 - A, z_0) \to Z_m (= \langle\varphi\rangle)$ by sending paths circling the odd indexed vertices into $\varphi$ and paths circling the even indexed vertices into $\varphi^{-1}$. The elements of $S_4$ permute the sets $A_1, A_2$ so that $S_4$ lifts to the Riemann surface $X_p$ covering $\mathbb{P}^1$ corresponding to the kernel of $\mu$. By the Riemann-Hurwitz formula

$$2p - 2 = -2m + 8(m - 1) \text{ or } p = 3(m - 1).$$

The order of the group on $X_p$ is $o(S_4)o(Z_m) = 24m$. Thus $X_p$ admits a group of order $8(p + 3)$.

q.e.d.

(The results of Theorems 7.2 and 7.3 together with the result that these bounds are sharp infinitely often were obtained independently by Colin Maclachlan [20] and the author [1]).

7.4. We shall soon see that the lower bound $8(p + 1) \le N(p)$ is sharp infinitely often, so the following definition is justified.

Definition. A group $G$ of automorphism on a Riemann surface $X_p$ will be called big if $o(G) \ge 8\ (p + 1)$.

In this section we shall show the existence of many sequences of big groups whose orders lie in arithmetic progressions. These groups will all be $(2,4,\lambda)$'s. So let us look at a $(2,4,\lambda) = \langle a,b \rangle$ where $a^2 = b^4 = c^\lambda = e$, $c = ab$, and the order of the group is $\mu\lambda$, where $\mu$ is the index of the cyclic group $\langle c \rangle$. Let $N$ be the smallest normal subgroup containing $b^2$. Then $\langle a,b \rangle / N$ can be expected to be a $(2,2,\lambda')$ where $\lambda' \mid \lambda$, a group of genus zero. If $N$ is abelian then it is an elementary abelian group of order $2^\ell$ for some $\ell$. Thus $\langle a,b \rangle$ is an extension by a dihedral group of an elementary abelian 2-group.

Theorem. (W. T. Kiley [16]). Let $\ell$ be a positive integer. If $p \equiv (1 - 2^\ell) \pmod{\ell 2^{\ell-1}}$ then $N(p) \ge 8(p - 1 + 2^\ell)$.

Proof. Fix positive integers $\ell$ and $m$. Let

$$A = \{\tau^k \mid \tau = e^{2\pi i/2\ell m},\ k = 0,1,\cdots, 2\ell m - 1\}$$

the vertices of a regular $2\ell m$-gon inscribed in the unit circle. Let $A_j = \{\tau^k \mid k \equiv j \pmod{\ell}$, $j = 0,1,\ldots,\ell-1$. There are $2m$ points in each set $A_j$. Let $D_{2\ell m}$ be generated by $R$ and $V$ where $Rz = \tau z$ and $Vz = z^{-1}$. Then $D_{2\ell m}$ permutes the sets $\{A_j\}$. Let $\gamma_k$ circle $\tau^k$ as before. Define $\mu : \pi_1(\mathbb{P}^1 - A, z_0) \to (Z_2)^\ell, (z_0 \notin A)$ where $(Z_2)^\ell = \langle \varphi_0, \varphi_1, \ldots \varphi_{\ell-1} \rangle$ by

$$\mu(\gamma_k) = \varphi_{\bar{k}} \quad \text{where } \bar{k} \equiv k \pmod{\ell}.$$

In $H_1(\mathbb{P}^1 - A, Z)$ $\ker\mu = \left\{\sum_{j=0}^{\ell-1} \sum_{k\equiv j} c_k \gamma_k \,\middle|\, \sum_{k\equiv j} c_k \equiv 0 \pmod 2\right\}$,

which is invariant under $D_{2\ell m}$.

The Riemann surface corresponding to $\ker\mu$, $X_p$, has a group of automorphisms of order $o(D_{2\ell m})\, o\big((Z_2)^\ell\big) = 4\ell m \cdot 2^\ell$.

By the Riemann-Hurwitz formula

$$2p - 2 = -2 \cdot 2^\ell + 2\ell m \cdot 2^{\ell-1}.$$

Thus $4 \cdot 2^\ell \cdot \ell m = 8(p - 1 + 2^\ell)$ and $p = 1 - 2^\ell + m \cdot \ell 2^{\ell-1}$ for all $\ell, m \geq 1$.

q.e.d.

For $\ell = 1$ we have the result of Theorem 7.2. For $\ell = 2$ we have: if $p \equiv 1 \pmod 4$ then $N(p) \geq 8(p+3)$ a result different from Theorem 7.3.

7.5. We now state a theorem which shows that the lower bounds on $N(p)$ discussed in the last three sections are sharp infinitely often.

Theorem 7.5. (W. T. Kiley [16]. Suppose we have positive integers $s$ and $t$ so that $s \geq 2$ and $4 \mid st$. Then there are an infinite number of genera $p$ so that

1) $N(p) \leq 8(p - 1 + s)$ and
2) $p \equiv (1 - s) \pmod{(st)/4}$.

(In Theorem 7.4 $s = 2^\ell$ and $t = 2\ell$, $\ell = 1,2,\ldots$.)

The proof of this theorem is an elaboration of the basic idea in Theorem 5.8 where Macbeath showed that $N(p) < 84(p-1)$ for infinitely many $p$. There $p = n + 1$ where $n$ was a large prime.

Here we shall assume $p - 1 = sn$ where again $n$ is a large prime satisfying further conditions. Moreover, a closer examination of the steps in the proof will allow us to actually compute $N(p)$ where $p = n + 1$, or $2n + 1$, where $n$ is prime. The proof of Theorem 7.5 will occupy the next several sections.

7.6. Some preliminary results.

From now on $n$ will denote a prime integer and $p$ will denote the genus of a Riemann surface $X_p$ where the subscript may be omitted if no confusion follows.

In the sequel we will need to recall Lemmas 5.5 and 5.7.

From elementary group theory we recall that the group of automorphisms of $Z_n$ is $Z_{n-1}$. Also by using the transfer one proves that if a finite group $G$ admits a central abelian subgroup $S$ of index $i$, then the transfer homomorphism from $G$ into $S$ is $g \to g^i$. If $S$ is a Sylow subgroup the map is surjective with kernel of order $i$ [27].

Other results from group theory will be proven in the propositions below.

Now we catalogue the big groups. If $X$ admits a big group of automorphism, $G$, then the genus of $X/G$ is zero (Lemma 5.5). Recalling the discussion of section 5.5 one sees that the possible big groups on a Riemann surface $X_p$ $(p \geq 2)$ are (for convenience we will divide them into 3 types):

| | | |
|---|---|---|
| Type (I) | $(2,5,\lambda)$ | $5 \leq \lambda \leq 19$ |
| | $(2,6,\lambda)$ | $6 \leq \lambda \leq 11$ |
| | $(2,7,\lambda)$ | $7 \leq \lambda \leq 9$ |
| | $(3,3,\lambda)$ | $4 \leq \lambda \leq 11$ |
| | $(3,4,\lambda)$ | $4 \leq \lambda \leq 5$ |
| | $(2,2,2,3)$ | |
| Type II | $(2,3,\lambda)$ | $\lambda \geq 7$ |
| Type III | $(2,4,\lambda)$ | $\lambda \geq 5$ |

To prove Theorem 7.5 we will put more and more conditions on the genus of a Riemann surface to eliminate big groups of Type I, Type II, and the unwanted big groups of Type III.

7.7. Theorem. Let $s$ be a positive integer and let $n$ be a prime. Let $p = sn + 1$. Let $G$ be a big group on $X_p$. Suppose

(i) $n > 84s$ ;
(ii) $n$ divides $o(G)$ .

Then $G$ admits a normal subgroup $H$ where $G/H$ is a cyclic group which is one of the following 4 possibilities:

(i) $(2,3,6)$ $G/H$ has order 6 and $X_p/H$ has genus one ;
(ii) $(2,4,4)$ $G/H$ has order 4 and $X_p/H$ has genus one ;
(iii) $(3,3,3)$ $G/H$ has order 3 and $X_p/H$ has genus one ;
(iv) $(2,5,10)$ $G/H$ has order 10 and $X_p/H$ has genus two.

If $3 \mid (n-1)$ then $o(G) \leq 24(p-1)$ .

If $3 \nmid (n-1)$ and $4 \mid (n-1)$ then $o(G) \leq 16(p-1)$ .

If $(12, n-1) = 2$ then $5 \mid (n-1)$ and $o(G) \leq 10(p-1)$ .

In particular, $\big((n-1)/2, 30\big) > 1$ .

Proof. Let $S_n$ be the n-Sylow subgroup of $G$ . Now $84(p-1) = 84sn \geq o(G)$ . Consequently $S_n$ is a normal cyclic subgroup, by Sylow's theorem. By Lemma 5.7 $X_p \to X_p/S_n$ is a smooth covering.

Consider the homomorphism $\Theta : G \to \mathrm{Aut}(S_n)$ given by $\Theta(g)(h) = ghg^{-1}$ . Let $H$ be the kernel of $\Theta$ . Then $S_n \subset H$ and $H$ is the centralizer of $S_n$ in $G$ . $S_n$ is thus a central subgroup of $H$ , and since $\mathrm{Aut}(S_n) \cong Z_{n-1}$ , $G/H$ is isomorphic to a subgroup of $Z_{n-1}$ . $G/H$ is a cyclic group whose order divides $n-1$ . The transfer map $H \to S_n$ is surjective and the kernel of the transfer, $N$ , is a normal complement to $S_n$ in $H$ ; that is, $H = NS_n$ where $N \cap S_n = \langle e \rangle$ . We thus have the following diagram.

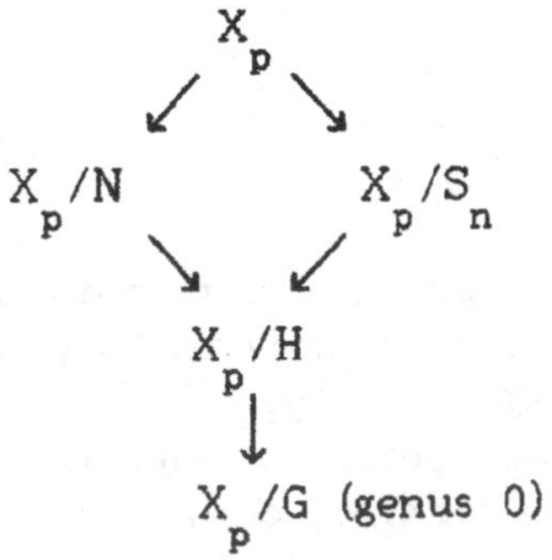

Now $H/N \cong S_n$ and $S_n$ acting on $X_p$ is without fixed points. Thus the covering $X_p/N \to X_p/H$ is also a smooth covering of $n$ sheets. Consequently, the genus of $X_p/H$ is positive. By Lemma 5.5, $o(H) \le 4(p-1)$. Since $G$ is big, $o(G/H) > 2$.

Now $G$ is a $(\nu_1, \nu_2, \dots, \nu_s)$ group $(s = 3$ or $4)$ and so $G/H$ is a $(\nu_1', \nu_2', \dots, \nu_s')$ group where $\nu_j' \mid \nu_j$. Since $G/H$ is cyclic and $X_p/H$ has positive genus, 3 of the $\gamma_i'$'s must be larger than one. By consulting the list of big groups one sees that the 4 possibilities for $G/H$ listed in the statement of the theorem are the only possibilities. Consequently $3, 4,$ or $5$ must be a factor of $n-1$.

If $3 \mid (n-1)$ then $G/H$ could be a $(2,3,6)$ of order $6$. Then $X_p/H$ has genus one so that $o(G) \le 6 \cdot 4(p-1)$. The remaining parts in the statement of the theorem are proved similarly. Note that if the genus of $X_p/H$ is 2 then $o(H) \le p-1$. (See the proof of Lemma 5.5.)

q.e.d.

<u>Corollary 7.7</u>. Let $s$ be a positive integer and let $n$ be a prime. Let $p = sn + 1$ and let $G$ be a big group on $X_p$. Suppose

(i) $n \ge 84s$

(ii) $\left(\frac{n-1}{2}, 30\right) = 1$ $\left(\text{i.e. } n \equiv 23, 47, \text{ or } 59 \pmod{60}\right)$.

Then $n \nmid o(G)$ and $G$ is not a big group of Type I.

<u>Proof</u>. That $n \nmid o(G)$ follows immediately from the theorem.

Now consider the 35 possibilities for the big groups of Type I. The

(2,2,2,3) is excluded since it has order $12(p-1)$ and $n \mid p-1$. The other 34 big groups are $(\alpha,\beta,\gamma)$'s. By the Riemann-Hurwitz formula $2p-2 = o(G)(1 - \frac{1}{\alpha} - \frac{1}{\beta} - \frac{1}{\gamma})$. We see that

$2p-2 = O(G) \cdot \frac{A}{B}$ where A and B are integers. Since $n \nmid o(G)$

it follows that $n \mid A$. By working out these 34 cases one sees that the largest prime dividing an A is 47. Since $n > 84$, this rules out all Type I big groups.

q.e.d.

7.8. The following group theoretic proposition will eliminate the $(2,3,\lambda)$'s.

Proposition 7.8. Let G be a non-cyclic $(2,3,\lambda)$ of order $\mu\lambda$.

Then $\lambda \le \mu^2$ and so $o(G) \le \mu^3$. In fact $\lambda \mid k(\lambda,\mu)$ where $k \le \mu$

and $k \mid \lambda$.

Proof. Suppose $G = \langle a,b \rangle$ where $a^2 = b^3 = c^\lambda = d^\lambda$, $c = ab$ and

$d = ba$. Since $dc = b^2$, $G = \langle c,d \rangle$. Now $\langle c \rangle \cap \langle d \rangle$ is normal, in fact, central in G. Consider the following diagram

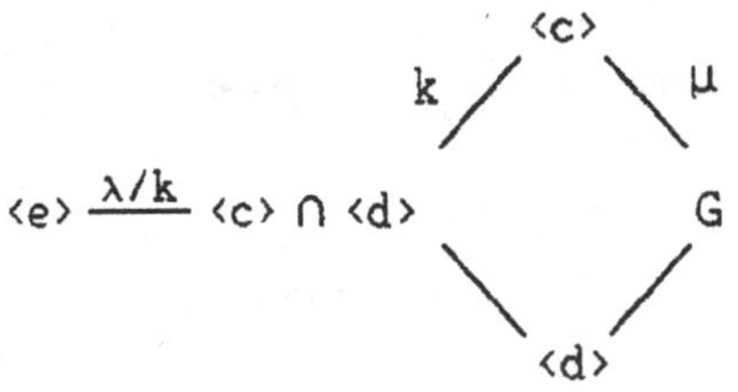


where $k\mu$ is the index of $\langle c \rangle \cap \langle d \rangle$ in G and $k \mid \lambda$.

Now $k \le \mu$. For in $G' = G/\langle c \rangle \cap \langle d \rangle = \langle c',d' \rangle$ the cyclic subgroups $\langle c'\rangle$ and $\langle d'\rangle$ intersect only in $\langle e'\rangle$ and so $c'^{\alpha}d'^{\beta}$ are distinct for $0 \le \alpha,\beta \le k-1$, $(\alpha,\beta) \ne (0,0)$. Thus $k^2 \le o(G') = k\mu$ Also $6 \mid k\mu$. If not, say $2 \nmid k\mu$. Then $a \in \langle c\rangle \cap \langle d\rangle \subset \langle c\rangle$ and so $G$ is cyclic. Similarly, $3 \mid k\mu$.

The transfer $t : G \to \langle c\rangle \cap \langle d\rangle$ is the map $g \to g^{k\mu}$. But $6 \mid k\mu$; therefore, a and b map into the identity; that is, $g \to g^{k\mu}$ is the trivial homomorphism. Consequently $c^{k\mu} = e$ or $\lambda \mid k\mu$. Thus $\lambda \le k\mu \le \mu^2$. That $\lambda \mid k(\lambda,\mu)$ is now immediate.

q.e.d.

<u>Theorem 7.8</u>. Assume the hypotheses of Corollary 7.7. Assume, moreover, that $n \ge 216s^2 = 6^3s^2$. Then G is a $(2,4,\lambda)$ group.

<u>Proof</u>. If G is a $(2,3,\lambda)$ of order $\mu\lambda$, then by the Riemann-Hurwitz formula

$$2ns = 2p - 2 = o(G)\left(1 - \frac{1}{2} - \frac{1}{3} - \frac{1}{\lambda}\right) = \mu\lambda\left(\frac{1}{6} - \frac{1}{\lambda}\right) = \frac{\mu}{6}(\lambda - 6)$$

or

$$12ns = \mu(\lambda - 6).$$

Since $n \nmid \mu$ we see that $n \mid (\lambda - 6)$ or $\lambda - 6 = \tau n$.

Thus

$$12s = \tau\mu$$

and

$$\mu = 12s/\tau \le 12s.$$

By the proposition $o(G) \le (12s)^3$.

But G is a big group. Therefore

$$o(G) \ge 8(p+1) > 8(p-1) = 8ns \ge (12s)^3,$$

a contradiction.

q.e.d.

7.9. Now we prove another group theoretic proposition, very similar to that in 7.8, to help eliminate the unwanted $(2,4,\lambda)$'s .

Proposition 7.9. Let $G$ be a $(2,4,\lambda)$ group of order $\mu\lambda$ where $\mu > 4$ , $4 \mid \mu$ and $\lambda > \mu^2$ $\left(o(G) > \mu^3\right)$.

Then there is a positive integer $k$ so that $3 \leq k \leq \mu/2$ and $k \mid \lambda$ . If $k = 3$ then $\mu = 8$ .

Proof. Let $G = \langle a,b \rangle$ where $a^2 = b^4 = c^\lambda = d^\lambda = e$ , $c = ab$ , $d = ba$ . We show first that $G \neq \langle c,d \rangle$ .

Suppose $G = \langle c,d \rangle$ . Since $4 \mid \mu$ it follows exactly as in Proposition 7.8 that if the index of $\langle c \rangle \cap \langle d \rangle$ is $k\mu$ , the transfer homomorphism $G \to \langle c \rangle \cap \langle d \rangle$ , $g \to g^{km}$ is the identity and consequently $\lambda \mid km$ . Also $k \leq \mu$ as before, so $\lambda \leq \mu^2$ . But this is excluded by hypothesis.

Now $\langle c,d \rangle$ is normal (since $aca = d$) and must now have index 2 in $G$ . Consider the diagram

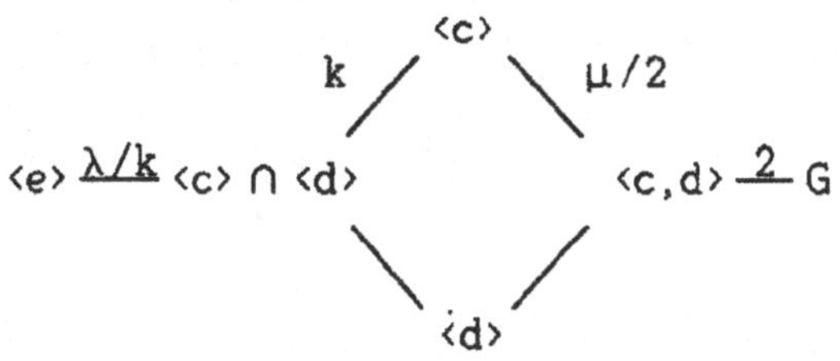


Then considering $\langle c,d \rangle / \langle c \rangle \cap \langle d \rangle = \langle c',d' \rangle$ we see as in Proposition 7.8 that $k \leq \mu/2$ .

If $k = 1$ $\langle c \rangle = \langle d \rangle$ and so $\mu = 2$ , a contradiction.

If $k = 2$ $G/\langle c\rangle \cap \langle d\rangle$ is a $(2,4,2)$, of order 8 and so $\mu = 4$, a contradiction.

If $k = 3$ $G/\langle c\rangle \cap \langle d\rangle$, is a $(2,4,3)$ of order 24 and so $\mu = 8$.

q.e.d.

<u>7.10</u>. Proof of Theorem 7.6

We assume $s$ and $t$ are positive integers so that $s \geq 2$ and $4 \mid st$. Let $c$ be a positive integer so that

(a) if $1 \leq k \leq 4s$ then $k \mid c$.

(b) $(st/4) \mid c$.

Let $n$ be a prime integer so that $n \equiv -1 \pmod{c}$ and $n > 216s^2$. Since $60 \mid c$ we see that $n \equiv 59 \pmod{60}$. Also $n > 84s$. Let $p = sn + 1$. Then

$$p = 1 - s + s(n+1) \equiv (1-s) \pmod{c} \equiv (1-s) \left(\mathrm{mod}\left(\tfrac{st}{4}\right)\right).$$

By Dirichet's theorem on primes in an arithmetic series, there are an infinite number of such genera $p$. It remains to show that

$$N(p) \leq 8(p - 1 + s).$$

Suppose $G$ is a big group on $X_p$. Thus by Corollary 7.7 and Theorem 7.8 $G$ is a $(2,4,\lambda)$ group of order, say $\mu\lambda$ and $n$ does not divide $o(G)$. By the Riemann-Hurwitz formúla

$$8p - 8 = \mu(\lambda - 4).$$

Now $n$ divides $p - 1$ but $n$ does not divide $o(G)$.

Therefore $n \mid (\lambda - 4)$. Let $\lambda - 4 = \tau n$. Then

$$8ns = 8(p-1) = \mu n \tau \quad \text{or} \quad \mu = 8s/\tau.$$

Since $o(G) = \mu\lambda = 8p - 8 + 4\mu$ it remains to show

$$\mu \leq 2s \quad \text{or} \quad \tau \geq 4.$$

We eliminate the other possibilities for $\tau$. Now

$$\lambda = n\tau + 4 = \tau(n+1) + (4-\tau).$$

Letting $\tau = 1, 2$ or $3$ we see that $4 - \tau = 3, 2$ or $1$.

We apply Proposition 7.9 to produce a $k$ so that

$3 \le k \le \mu/2 (= 4s/\tau)$ and $k \mid \lambda$. If $k = 3$ then $\mu = 8$. By the choice of $n$ we also have $k \mid (n+1)$. Thus $k$ divides $1, 2$ or $3$, all cases which are impossible $\left(k = 3 \Rightarrow \tau = 1 \Rightarrow \mu \ge 16\right)$.

Thus $\tau \ge 4$.

q.e.d.

7.11. In order to apply the proof of Theorem 7.5 to produce genera $p$ where $N(p)$ is known, applying condition (a) in Section 7.10 together with $n > 216s^2$ quickly leads to rather large numbers. We conclude this chapter by proving several theorems which allow us to compute $N(p)$ for some genera which are relatively small.

Theorem 7.11. Let $n$ be a prime integer, $n \ge 84$ $(p - 1 = n)$.

(a) If $3 \mid (n-1)$ then $N(n+1) = 12n$;

i.e. $N(p) = 12(p-1)$.

(b) If $3 \nmid (n-1)$ and $5 \mid (n-1)$, then $N(p) = 10n$;

i.e. $N(p) = 10(p-1)$.

(c) If $(n-1, 15) = 1$, then $N(n+1) = 8(n+4)$;

i.e. $N(p) = 8(p+3)$.

The proof of this theorem is derived by examining more closely Theorem 7.7 and 7.8 when $s = 1$.

7.12. Theorem. Let $n$ be a prime integer, $n > 84$, $p = n + 1$.

If $3 \mid (n-1)$, then $N(p) \le 12(p-1)$.

If $3 \nmid (n-1)$ and $5 \mid (n-1)$, then $N(p) \le 10(p-1)$.

If $(n-1, 15) = 1$ and $G$ is a big group, then $G$ is not of Type I and $n \nmid o(G)$.

<u>Proof</u>. Again consider the diagram in the proof of Theorem 7.7 when assuming that $n \mid o(G)$ where $G$ is a big group.

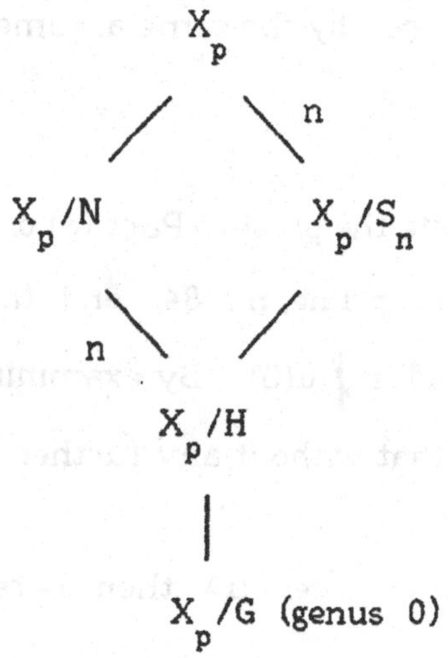


Since $X_p \to X_p/S_n$ is smooth we see that the genus of $X_p/S_n$ is 2. The genus of $X_p/H$ is one or 2. A Galois cover $X_2 \to X_1$ must have precisely 2 sheets so the order of $H$ is $n$ or $2n$. If the genus of $X_p/H$ is one then $G/H$ is a cyclic group of genus one and therefore has order bounded by 6, and the order divides $n-1$. Thus if $3 \mid (n-1)$ the maximum order for $G$ is $(6)(2n) = 12n$. (If $o(G) = 12n$ one sees that $G$ must be a $(2,6,6)$).

If $3 \nmid (n-1)$ and the genus of $X_p/H$ is one, the maximum order for $G/H$ is 4 and $o(G) \le 8n$ which is not big. So if $3 \nmid n-1$ and $G$ is big then the genus of $X_p/H$ is 2. In this case $G/H$ must be a cyclic group of genus 2, a $(2,5,10)$ which has order 10. Thus if $3 \nmid (n-1)$ and $5 \mid (n-1)$ then the maximum order of $G$ is $10n$. If $3 \nmid (n-1)$ and $5 \nmid (n-1)$ then the situation illustrated in the diagram is impossible and $n \nmid o(G)$. Then we eliminate groups of Type I by the same argument as in the proof of Corollary 7.7.

q.e.d.

7.13. We now complete the proof of Part (c) of Theorem 7.11. We assume $p = n+1$, n is prime, $n \ge 84$, and $(n-1,15) = 1$. $X_p$ admits a big group and $n \nmid o(G)$. By examining the proofs in Section 7.8 we show that without any further restrictions $G$ cannot be a $(2,3,\lambda)$.

If $G$ is a $(2,3,\lambda)$ of order $\mu\lambda$ then there is a $k$ so that $k \mid \lambda$, $k \le \mu$ and $\lambda \mid k(\lambda,\mu)$. Since $p-1 = n$, the Riemann-Hurwitz formula gives $2p-2 = \mu\lambda(\frac{1}{6} - \frac{1}{\lambda})$ or

$$12n = \mu(\lambda - 6).$$

As before $n \mid \lambda - 6$ so that $\lambda - 6 = \tau n$ and $12 = \mu\tau$. The possibilities for $\tau$ are $1,2,3,4$ and $6$ and correspondingly $\mu = 12,6,4,3$ and $2$.

If $\tau = 1$, $\mu = 12$, $\lambda = n+6$ then $(\mu,\lambda) = 1$. Thus $\lambda \mid k$; that is $\lambda = k \le \mu = 12$. Consequently $n + 6 \le 12$. This contradicts $n \ge 84$. Thus $\tau$ cannot be 1.

If $\tau = 2$ , $\mu = 6$ , $\lambda = 2n + 6$ , then $(\mu,\lambda) = 2$ . Thus $\lambda \mid 2k$ and so $2n + 6 \le 12$ , again a contradiction .

For $\tau \ge 3$ , $\mu \le 4$ so directly by Proposition 7.8 $o(G) \le \mu^3 \le 64$ . Thus our big group is not a $(2,3,\lambda)$ .

We conclude that $G$ must be a $(2,4,\lambda)$ group of order $\mu\lambda$ . Again the Riemann-Hurwitz formula gives

$$8n = \mu(\lambda - 4) .$$

Again $\lambda - 4 = \tau n$ and $\mu\tau = 8$ . Therefore $o(G) = \mu\lambda = \frac{8}{\tau}(\tau n + 4) \le 8(n + 4) = 8(p + 3)$ .

But if $(n - 1,15) = 1$ then $n = 2,8$ or $14 \pmod{15}$ . Consequently $p = n + 1 \equiv 0 \pmod 3$ . By Theorem 7.3 $N(p) = 8(p + 3)$ .

The proof of Part (c) of Theorem 7.11 is complete.

7.14. To complete the proof of Parts (a) and (b) of Theorem 7.11 we first exhibit two groups of genus 2 which must act on $X_p/S_n$ in the diagram of Section 7.12.

The $(2,6,6)$ is $Z_2 \times Z_6 = \langle a,b \rangle$ where $a^2 = b^6 = e$ .

The $(2,5,10)$ is $Z_{10} = \langle \varphi^5,\varphi^2 \rangle$ where $\varphi$ generates $Z_{10}$ .

If $3 \mid (n - 1)$ ($n$ prime) we want to extent $Z_n$ by the above $(2,6,6)$ to obtain a $(2,6,6)$ of order $12n$ . This will complete the proof of Part (a) of Theorem 7.11.

If $5 \mid (n - 1)$ ($n$ prime) we want to extend $Z_n$ by the above $(2,5,10)$ to obtain a $(2,5,10)$ of order $10n$ . This will complete the proof of Part (b) of Theorem 7.11.

<u>Proposition 7.14</u> Let $G$ be an $(\alpha,\beta,\gamma)$ where the l.c.m. $[\alpha,\beta]$ divides $\gamma$. Let $G = \langle a,b \rangle$ where $a^\alpha = b^\beta = c^\gamma = e$ $(c = ab)$. Assume $\alpha,\beta,\gamma \geq 2$ and at most one of these integers is $2$.

(i) Suppose there exists a surjective homomorphism

$$\Theta : G \to Z_{[\alpha,\beta]}$$

so that $o\big(\Theta(a)\big) = \alpha$ and $o\big(\Theta(b)\big) = \beta$.

(ii) Suppose $n$ is prime and $[\alpha,\beta] \,\big|\, (n-1)$ but $n \nmid o(G)$.

Then there exists $G'$, again an $(\alpha,\beta,\gamma)$ of order $o(G)n$, and $G'$ satisfies hypothesis (i).

<u>Proof</u>. $\Theta$ maps $G$ onto $Z_{[\alpha,\beta]}$ which is a subgroup of $\mathrm{Aut}(Z_n)$. Let $G'$ be the semi-direct product of $G$ and $Z_n$. That is, $G' = \langle a,b,\varphi \rangle$ where $G = \langle a,b \rangle$, $\langle \varphi \rangle = Z_n$ and $\langle a,b \rangle \cap \langle \varphi \rangle = \langle e \rangle$, together with the following relations:

$$a\varphi a^{-1} = \varphi^x,\ x^\alpha \equiv 1 \pmod{n},\ x \not\equiv 1 \pmod{n};$$

$$b\varphi b^{-1} = \varphi^y,\ y^\beta \equiv 1 \pmod{n},\ y \not\equiv 1 \pmod{n}.$$

Thus $c\varphi c^{-1} = \varphi^{xy}$ and $(xy)^{[\alpha,\beta]} \equiv 1 \pmod{n}$.
$x$ and $y$ can be chosen so that $xy \not\equiv 1 \pmod{n}$ since $\alpha$ and $\beta$ are not both $2$.

Now $(a\varphi)^t = (a\varphi a^{-1})(a^2\varphi a^{-2})\cdots(a^t\varphi a^{-t})a^t = \varphi^{x+x^2+\cdots x^t}a^t$.

Consequently $(a\varphi)^\alpha = \varphi^{x+x^2+\cdots x^{\alpha-1}+1}a^\alpha = e$

$$(b\varphi)^\beta = e$$

and $\quad (c\varphi)^{[\alpha,\beta]} = c^{[\alpha,\beta]}$ $\big($since $xy \not\equiv 1 \pmod{n}\big)$.

Thus $\quad (c\varphi)^\gamma = e$.

Now $\langle a, b\varphi\rangle$, a subgroup of $G'$, is an $(\alpha, \beta, \gamma)$. To show that $G' = \langle a, b\varphi\rangle$ we need only show $\varphi \in \langle a, b\varphi\rangle$. Consider the commutator $[a, b\varphi] \in \langle a, b\varphi\rangle$.

$$[a, b\varphi] = c\varphi a^{-1}(b\varphi)^{-1}$$

$$= \varphi^{xy} c a^{-1} b^{-1} \varphi^{-y}$$

$$= \varphi^{y(x-1)}[a, b]$$

$$[a, b\varphi]^{o(G)} = \varphi^{y(x-1)o(G)}$$

Now $y(x-1)o(G) \not\equiv 0 \pmod{n}$, so $\varphi \in \langle a, b\varphi\rangle$.

Since $\langle\varphi\rangle$ is normal in $G'$ and $G'/\langle\varphi\rangle \cong G$, the proof of the proposition is easily completed.

q.e.d.

The proof of Theorem 7.11 is completed by applying the proposition to the two cases discussed above, $(2,6,6)$ and $(2,5,10)$.

<u>7.15</u>. We now show that the two other bounds given in Theorem 7.7 are sharp infinitely often, by letting $s = 2$ in the theorem.

<u>Theorem 7.15</u>. Suppose $n$ is a prime integer, $n \geq 168$. Let $p = 2n + 1$.

If $3 \mid (n-1)$ then $N(p) = 24(p-1)$.

If $3 \nmid (n-1)$ and $4 \mid (n-1)$ then $N(p) = 16(p-1)$.

<u>Proof</u>. It suffices to show the existence of two groups satisfying the hypotheses of Proposition 7.14, namely:

(i) a $(2,3,12)$ of order 48 and genus 3 with a $(2,3,6)$ of order 6 as a homomorphic image.

(ii) a $(2,4,8)$ of order 32 and genus 3 with a $(2,4,4)$ of order 4 as a homomorphic image.

(i) Let $A_4 = (2,3,3)$ be represented as the symmetries of the regular tetrahedron inscribed in the Riemann sphere. Let

$B = \{b_1, b_2, b_3 b_4\}$ be the 4 vertices which are fixed points for elements of order 3 in $A_4$. Define $\mu : \pi_1(\mathbb{P}^1 - B, b) \to Z_4$ $(= \langle\varphi\rangle)$ by $\mu(\gamma_i) = \varphi$ for each of the 4 paths circling the $b_i$'s. The kernel of $\mu$ is associated to a 3-sheeted cover of $\mathbb{P}'$ with total ramification 12, that is, a Riemann surface $X_3$. $A_4$ lifts to $X_3$ and so $X_3$ admits a $(2,3,12)$ of order 48. If the group is $G = \langle a,b\rangle$, $a^2 = b^3 = (ab)^{12} = e$, it is not difficult to see that the commutator subgroup of $G$ is $\langle abab^2, ab^2ab\rangle$ $(= G')$ and $G'$ is isomorphic to the quaternion group of order 8. Thus $G/G' \cong Z_6$ and $G/G'$ is a $(2,3,6)$.

(ii) Let $Z_4 \times Z_4 = \langle\varphi,\psi\rangle$. Let $a$ be the automorphism of $Z_4 \times Z_4$ of order 2 that permutes $\varphi$ and $\psi$. Let $G = \langle a,\varphi,\psi\rangle$ be the semi-direct product where $a^2 = \varphi^4 = \psi^4 = e$, $\varphi\psi = \psi\varphi$, $a\varphi a = \psi$ and $a\psi a = \varphi$. It follows that $G = \langle a,\varphi\rangle$, a $(2,4,8)$ of order 32. The subgroup $\langle\varphi\psi^3\rangle$ is seen to be normal and

$$G/\langle\varphi\psi^3\rangle \cong Z_2 \times Z_4 \text{, a } (2,4,4).$$

Thus $G/\langle\varphi\psi^3, a\varphi^2\rangle \cong Z_4$, a $(2,4,4)$.

q.e.d.

All the Riemann surfaces occuring in Theorem 7.15 are elliptic-hyperelliptic. If $G = \langle a,b\rangle$, $c = ab$ then $c^6$ (resp. $c^4$)

is the elliptic-hyperelliptic involution in the case of $(2,3,12)$ $\big($resp. $(2,4,8)\big)$.

7.16. We continue this discussion of $N(p)$ by showing that the results of Theorem 7.11 hold for any prime integer $n \geq 11$, $n \neq 13$. Macbeath [19] showed that $N(14) = (13)(84)$.

The following remarks will be useful. If a Riemann surface of genus $p$ admits an automorphism group of order $m(p-1)$ and $m \geq 24$, then the only possible values for $m$ are 24, (132/5), 30, 36, 40, 48, and 84. The reader will see that part of the following discussion depends on looking fairly carefully at the 35 Type I groups. This is slightly tedious but quite easy. Also a useful table of non-cyclic simple groups of low order is appended to this chapter.

Lemma 7.16. Let $n$ be a prime integer, $n \geq 11$, $n \neq 13$ and $n < 84$. Let $p = n + 1$. Let $G$ be a big group on $W_p$ and suppose $n \mid o(G)$. Then $n^2 \nmid o(G)$, $G$ is solvable, and the $n$-Sylow subgroup is normal in $G$.

Proof. The only big automorphism group of order $m(p-1)$ where $m$ is prime is a $(2, 3, 78)$ of order $13(p-1)$. Consequently, $n^2 \nmid o(G)$.

Suppose $G$ is not solvable. Then $G$ has a non-cyclic simple factor in any composition series. If $o(G) = mn$, we may assume $m \leq 84$, and so $o(G) < (84)^2$ $(= 7056)$. By considering the non-cyclic simple groups of order less than 7056, we see that the relevant prime factors of the orders of these simple groups (11, 17, 19, 23) have indices (e.g. 60, 144, 180, etc.) inappropriate for automorphism groups. We conclude that $G$ is solvable.

Since $G$ is solvable, in any composition series there is a group $H$ with a subgroup $N$, normal in $H$, so that $H/N = Z_n$. Consider the coverings:

$$W_p \to W_p/N \to W_p/H .$$

The covering $W_p/N \to W_p/H$ is an unramified $Z_n$ covering, and so the genus of $W_p/H$ is positive. Consequently, $o(H) \le 4(p-1) = 4n$. Therefore, $o(N) \le 4$ and $S_n$ (the n-Sylow subgroup) is normal in $H (n \ge 11)$. $S_n$ is a characteristic subgroup of H and so is normal in the next subgroup of the composition series and so on until we reach G.

q.e.d.

Now that we know that $S_n$ is normal, the proof of Theorem 7.11 proceeds exactly as before until last sentence, where we consider Type I groups, G , with $n \nmid o(G)$ . Such a group has order $(B/A)(2p-2) = (2Bn/A)$ where n must divide A. The possible values for A are 11, 17, 19, 23, 29, 31, 41, and 47, and so these are the values of n to consider. In all but two cases the $(\alpha , \beta , \gamma)$ Type I group is perfect $\big((\alpha , \beta) = (\beta , \gamma) = (\gamma ,\alpha) = 1\big)$ and so must have a non-cyclic simple group is any composition series. We then see that, with one exception, o(G) is not divisible by the order of a non-cyclic simple group and so cannot occur. The exceptional perfect group is a (2, 5, 9) group of order (180/17). However, Paul Hewitt showed the author that there does not exist a (2, 5, 9) group of order 180. The two non-perfect Type I candidates fail because the group would be too small. For example, a (2, 5, 16) group of order (160/19) on $W_{20}$ would yield a group of order 160 which is less than $8(p+1)$ (= 168). Thus Type I groups are eliminated from the competition to achieve $N(p)$.

An examination of the proofs in Section 7.13 shows that they work equally well for the primes now under consideration and so Theorem 7.11 does indeed hold for all primes $n \geq 11$, $n \neq 13$. $p = 12$ is the smallest genus for which $N(p) = 8(p+3)$.

7.17. Now we consider $N(p)$ for $p = 2n+1$, $n$ prime and $n \geq 11$. We will prove the following.

Theorem 7.17. Let $n$ be a prime integer greater than or equal to 11. Let $p = 2n+1$.

a) If $3 \mid (n-1)$ then $N(p) = 24(p-1)$

b) If $3 \nmid (n-1)$ and $4 \mid (n-1)$ then $N(p) = 16(p-1)$

c) If $((n-1), 12) = 1$ then $N(p) = 8(p+1)$

Remarks. $p = 23$ is the smallest genus for which $N(p) = 8(p+1)$. Paul Hewitt showed this by finding all previously unknown $N(p)$ for $p \leq 25$. Notice that we have lower bounds in parts a) and b) of the theorem by Theorem 7.15.

The integers $n$ and $p$ will always have the above meaning in this section.

We now adapt Theorem 7.7 to the present situation.

Lemma 7.17. Suppose $o(G) = N(p)$ and $n \mid o(G)$. Then

a) $n^2 \nmid o(G)$.

b) For $n < 168$, $G$ is solvable.

c) For all $n$, $S_n$ (the $n$-Sylow subgroup) is normal in $G$.

Proof. $o(G) = m(p-1)$. The only case where $m$ is prime is a $(2, 3, 78)$ of order $13(p-1)$. This is excluded for $p = 23$ since 3 does not divide the order of the group. Thus $n^2 \nmid o(G)$ .

If $o(G) = m(2n)$ then $m \le 84$ , and $o(G) \le 168n$. If $n > 168$ then $S_n$ is normal.

If $n < 168$ then $o(G) < (168)^2$ (= 28, 224). But all non-cyclic simple groups of order less than 28, 224 have the relevant cyclic Sylow subgroups with too great an index to be an automorphism group of a Riemann surface, with the possible exception of a simple group of order 660 (= 22.30)

We consider the possibility of a (2, 3, 10) on $W_{23} \cdot S_{11}$ in this group is unramified and by Sylow's theorem has index 5 in its normalizer. Thus $W_{23}/S_{11}$ admits an automorphism of order 5. We now reach the contradiction by observing that the genus of $W_{23}/S_{11}$ is 3.

Thus G cannot be simple. In fact, it must be solvable since a non-cyclic simple group cannot be fitted into a composition series. We now use the argument in Lemma 7.15 to show that $S_n$ is normal.

q.e.d.

We can now apply the argument of Theorem 7.7 which results from $S_n$ being normal in G. In the figure accompanying the proof of that theorem $W_p/S_n$ has genus 3 . Consequently, possibility (iv) in the statement of the theorem is ruled out because, again, a $W_3$ does not admit a $Z_5$ . Thus in the light of Theorem 7.15, the conclusion of Theorem 7.7 for our n and p is:

If $3 \mid (n-1)$ then $o(G) = 24(p-1)$.

If $3 \nmid (n-1)$ and $4 \mid (n-1)$ then $o(G) = 16(p-1)$.

In any case, $(n-1, 12) \neq 1$.

Now Corollary 7.7 reads: If $G$ is big and $(n-1, 12) = 1$ then $n \nmid o(G)$.

We now eliminate Type I groups by observing that their orders lose out in the competition or they are impossible because they are perfect.

To eliminate Type II groups we proceed as in Proposition 7.8. $G$ is a $(2, 3, \lambda)$ of order $\mu\lambda$ where $n \nmid o(G)$. $2p-2 = (\mu/6)(\lambda-6)$. Consequently, there is an integer $\tau$ which divides $24$ and $\lambda = \tau n + 6$, and $\mu = 24/\tau$. Finally there is an integer $k$ dividing $\lambda$ so that $\lambda \mid k(\lambda, \mu)$ and $k \leq \mu$.

If $\tau = 1$ we see that $(\lambda, \mu) = 1$ since $n$ is prime, and so $\lambda = k$. It follows that $n < 18$ and so $\lambda$ has the possible values 17, 19 and 23. Since all these groups would be perfect, they are eliminated.

If $\tau = 2$ we see that $(\lambda, \mu) = 4$, and so $\lambda \leq 48$, or $n < 21$. The possible values for $n$ are 19, 17, 13 and 11, and the possible values for $\lambda$ are 44, 40, 32 and 28. The first three possibilities lead to groups that are too small.

Thus we are led to consider the possibility of a $(2, 3, 28)$ of order 336 on a $W_{23}$. If we denote $G$ by $\langle a, b \rangle$ with $c = ab$ and $d = ba$, then $G = \langle c, d \rangle$ and the normal subgroup $\langle c \rangle \cap \langle d \rangle$ must have order 4. The quotient of $G$ by this normal subgroup would then have to be a $(2, 3, 7)$ of order 84, a contradiction.

The larger values of $\tau$ are easily eliminated and we see that our big group must be a $(2, 4, \lambda)$. For these groups we have:

$$2p - 2 = 4n = \mu(\lambda - 4)/4 \text{ or } 16n = \mu(\lambda - 4).$$

Since $n \nmid o(G)$ we see that there is an integer $\tau$ so that $\lambda - 4 = \tau n$, and so $\lambda = \tau n + 4$, $\mu = 16/\tau$ and $o(G) = 8\left(2n + (8/\tau)\right)$. The proof will be completed by showing that $\tau = 1$ or $2$ is impossible.

We let $G = \langle a, b\rangle$, $c = ab$ and $d = ba$, as before. We must first show that $G \neq \langle c, d\rangle$. Assuming the contrary, we use the transfer onto the central subgroup $\langle c\rangle \cap \langle d\rangle$ to find an integer $k$, dividing $\lambda$, $k \le \mu$, so that $\lambda \mid \mu k$. $\mu$ is always a power of 2. If $\tau = 1$, $\lambda$ is odd. Now $G$ is simultaneouly a $(2, 4, \lambda)$ and a $(\lambda, \lambda, 2)$. Thus the abelianized $G$ must be cyclic with order dividing 2 and $\lambda$. Thus we reach the contradiction that $G$ is perfect. If $\tau = 2$, we see that $(\lambda, \mu) = 2$, $\lambda \le 16$ and so $n$ is too small. We conclude that the index of $\langle c.d\rangle$ in $G$ is 2.

Now refer to the diagram accompanying Proposition 7.9.

First assume $\tau = 2$. $\lambda = 2n + 4$, and so $\lambda$ is not divisible by 4. Consequently, $k = 2$ or 3. But if $k = 2$, $G/\langle c\rangle \cap \langle d\rangle$ is a $(2, 4, 2)$ of order 16, a contradiction. If $k = 3$ then $3 \mid (2n + 4)$ which implies that $3 \mid (n - 1)$, in which case the Type III groups are not in the competition.

Now assume $\tau = 1$, $\mu = 16$, $\lambda = n + 4$, and $k \le 8$. Since $n$ is prime, $k$ must be odd. Let $G' = G/\langle c\rangle \cap \langle d\rangle$. $k = 3$ implies that $G'$ is a $(2, 4, 3)$ of order 48, a contradiction. $k = 5$ implies that $G'$ is a $(2, 4, 5)$ of order 80, again a contradiction since it implies that a $W_3$ admits an automorphism of order 5. If $k = 7$, we have a $(2, 4, 7)$

of order 112. Let $c'$ and $d'$ be the elements in $G'$ corresponding to $c$ and $d$ in $G$. $o(\langle c', d'\rangle)$ is 56, and since $S_7$ is not normal in $\langle c', d'\rangle$, $S_2$ is. $o(S_2) = 8$ and being characteristic in $\langle c', d'\rangle$, it is normal in $G'$. $S_2$ has as non-identity elements

$$\left\{c'^{-k} b'^{2} c'^{k} \;\middle|\; k = 0, 1, \ldots, 6\right\},$$

and so $S_2$ is seen to be an elementary abelian group of order 8. $G'/S_2$ acts faithfully on $S_2$ as a dihedral group of order 14. This is the final contradiction since the automorphisms of $S_2$ form the simple group of order 168, a group which does not have as a subgroup a dihedral group of order 14.

This completes the proof of Theorem 7.17.

With the work of Paul Hewitt we know $N(p)$ for all $p = n + 1$, or $2n + 1$ where $n$ is a prime integer.

Orders of non-cyclic simple groups $< (168)^2 = 28224$

$60 = 2^2.3.5$

$168 = 2^3.3.7$

$360 = 2^3.3^2.5$

$504 = 2^3.3^2.7$

$660 = 2^2.3.5.11$

$1092 = 2^2.3.7.13$

$2448 = 2^4.3^2.17$

$2520 = 2^3.3^2.5.7$

$3420 = 2^2.3^2.5.19$

$4080 = 2^4.3.5.17$

$5616 = 2^4.3^3.13$

$6048 = 2^5.3^3.7$

$$6072 = 2^{3}.3.11.23$$
$$7800 = 2^{3}.3.5^{2}.13$$
$$7920 = 2^{4}.3^{2}.5.11$$
$$9828 = 2^{2}.3^{3}.7.13$$
$$12180 = 2^{2}.3.5.7.29$$
$$14880 = 2^{5}.3.5.31$$
$$20160 = 2^{6}.3^{2}.5.7$$
$$25308 = 2^{2}.3^{2}.19.37$$
$$25920 = 2^{6}.3^{4}.5$$

## REFERENCES


[1] Accola, R. D. M.; "On the number of automorphisms of a closed Riemann surface"; Transactions of the American Mathematical Society, Vol. 131 (1968), pp. 398-408.

[2] Accola, R. D. M.; "Riemann surfaces with automorphism groups admitting partitions"; Proceedings of the American Mathematical Society, Vol. 21 (1969), pp. 477-482.

[3] Accola, R. D. M.; "On generalized Weierstrass points on Riemann surfaces"; Modular Functions in Analysis and Number Theory; University of Pittsburg, Pittsburg, Pennsylvania (1978), pp. 1-19.

[4] Ahlfors, L. V. and Sario, Leo; Riemann Surfaces; Princeton University Press, 1960.

[5] Arbarello, E., Cornalba, M., Griffiths, P. A., Harris, J.; Geometry of Algebraic Curves, Vol. 1, Springer-Verlag, New York, 1984.

[6] Castelnuovo, G.; "Alcune osservazion sopra le serie irrazionali di gruppi di punti appartenenti ad una curva algebrica"; Rendiconti della R. Accademia dei Lincei, Series 4, VII, 1891 (Memorie scelte p. 71).

[7] Castelnuovo, G.; "Sui multiple di una serie lineare di gruppe di punti appartenente ad una curve algebrica"; Rendiconti del Circolo Matematico di Palermo, VII (1893), pp. 89-110 (Memorie scelte p. 95).

[8] Castelnuovo, G.; "Sulle serie algebriche di gruppi di punti appartenenti ad una curve algebrica"; Rendiconti della R. Accademia dei Lincei Series 5, XV, 1906 (Memorie scelte p. 509).

[9] Coolidge, J. L.; A Treatise on Algebraic Plane Curves; New York, Dover Publications, 1959.

[10] Enriques, F.; "Sopra le superficie che posseggono un fascio ellittico o di genere due di curve razionali"; Roma, Reale Accademia dei Lincei, Rendiconti (5), $7_2$, (1898), pp. 281-286.

[11] Farkas, H. M. and Kra, I.; Riemann Surfaces; New York, Heidelberg, Berlin, Springer-Verlag, 1980.

[12] Forster, O.; Lectures on Riemann Surfaces; New York, Heidelberg, Berlin, Springer-Verlag, 1981.

[13] Griffiths, P. and Harris, J.; Principles of Algebraic Geometry, New York, John Wiley and Sons, 1978.

[14] Gunning, R. C.; Lectures on Riemann Surfaces; Princeton Mathematical Notes, 1966.

[15] Hurwitz, A.; "Uber algebraische Gebilde mit eindeutigen Transformationen in sich"; Mathematischen Annalen (Vol. 41) (1893), pp. 403-442.

[16] Kiley, W. T.; "Automorphism groups on compact Riemann surfaces"; Transactions of the American Mathematical Society, Vol. 150 (1970), pp. 557-563.

[17] Lewittes, J.; "Automorphisms of compact Riemann surfaces"; American Journal of Mathematics, Vol. 85 (1963), pp. 734-752.

[18] Macbeath, A. M.; "On a theorem of Hurwitz"; Proceedings of the Glasgow Mathematical Association, Vol. 5 (1961), pp. 90-96.

[19] Macbeath, A. M.; Fuchsian Groups; Queen's College Dundee, University of St. Andrews, 1961.

[20] Maclachlan, C.; "A bound for the number of automorphism of a compact Riemann surface"; Journal of the London Mathematical Society, Vol. 44 (1969), pp. 265-272.

[21] Miller, G. A.; "Groups defined by the order of two generators and the order of their product"; American Journal of Mathematics, Vol. 24 (1902), pp. 96-100.

[22] Olsen, B. A.; "On higher order Weierstrass points"; Annals of Mathematics (2), Vol 95 (1972), pp. 357-364.

[23] Schwarz, H. A.; Gesammelte Mathematische Abhandlungen, Springer-Verlag, Vols. I and II (1890), Berlin.

[24] Segre, C.; "Intorno ai punti di Weierstrass"; Atti della Reale Academia dei Lincei, Series 5, Vol. viii (1899), pp. 89-91.

[25] Seifert, H. and Threlfall, W.; Lehrbuch der Topologie, Chelsea, New York, 1945.

[26] Walker, R. J.; Algebraic Curves, New York, Dover Publications, Inc., 1962.

[27] Zassenhaus, H.; The theory of groups, Chelsea, New York, 1949.

## Index

# Lecture Notes in Mathematics

For information about Vols. 1–1414
please contact your bookseller or Springer-Verlag

Vol. 1415: M. Langevin, M. Waldschmidt (Eds.), Cinquante Ans de Polynômes. Fifty Years of Polynomials. Proceedings, 1988. IX, 235 pages. 1990.

Vol. 1416: C. Albert (Ed.), Géométrie Symplectique et Mécanique. Proceedings, 1988. V, 289 pages. 1990.

Vol. 1417: A.J. Sommese, A. Biancofiore, E.L. Livorni (Eds.), Algebraic Geometry. Proceedings, 1988. V, 320 pages. 1990.

Vol. 1418: M. Mimura (Ed. ), Homotopy Theory and Related Topics. Proceedings, 1988. V, 241 pages. 1990.

Vol. 1419: P.S. Bullen, P.Y. Lee, J.L. Mawhin, P. Muldowney, W.F. Pfeffer (Eds.), New Integrals. Proceedings, 1988. V, 202 pages. 1990.

Vol. 1420: M. Galbiati, A. Tognoli (Eds.), Real Analytic Geometry. Proceedings, 1988. IV, 366 pages. 1990.

Vol. 1421: H.A. Biagioni, A Nonlinear Theory of Generalized Functions, XII, 214 pages. 1990.

Vol. 1422: V. Villani (Ed.), Complex Geometry and Analysis. Proceedings, 1988. V, 109 pages. 1990.

Vol. 1423: S.O. Kochman, Stable Homotopy Groups of Spheres: A Computer-Assisted Approach. VIII, 330 pages. 1990.

Vol. 1424: F.E. Burstall, J.H. Rawnsley, Twistor Theory for Riemannian Symmetric Spaces. III, 112 pages. 1990.

Vol. 1425: R.A. Piccinini (Ed.), Groups of Self-Equivalences and Related Topics. Proceedings, 1988. V, 214 pages. 1990.

Vol. 1426: J. Azéma, P.A. Meyer, M. Yor (Eds.), Séminaire de Probabilités XXIV, 1988/89. V, 490 pages. 1990.

Vol. 1427: A. Ancona, D. Geman, N. Ikeda, École d'Eté de Probabilités de Saint Flour XVIII, 1988. Ed.: P.L. Hennequin. VII, 330 pages. 1990.

Vol. 1428: K. Erdmann, Blocks of Tame Representation Type and Related Algebras. XV. 312 pages. 1990.

Vol. 1429: S. Homer, A. Nerode, R.A. Platek, G.E. Sacks, A. Scedrov, Logic and Computer Science. Seminar, 1988. Editor: P. Odifreddi. V, 162 pages. 1990.

Vol. 1430: W. Bruns, A. Simis (Eds.), Commutative Algebra. Proceedings. 1988. V, 160 pages. 1990.

Vol. 1431: J.G. Heywood, K. Masuda, R. Rautmann, V.A. Solonnikov (Eds.), The Navier-Stokes Equations – Theory and Numerical Methods. Proceedings, 1988. VII, 238 pages. 1990.

Vol. 1432: K. Ambos-Spies, G.H. Müller, G.E. Sacks (Eds.), Recursion Theory Week. Proceedings, 1989. VI, 393 pages. 1990.

Vol. 1433: S. Lang, W. Cherry, Topics in Nevanlinna Theory. II, 174 pages. 1990.

Vol. 1434: K. Nagasaka, E. Fouvry (Eds.), Analytic Number Theory. Proceedings, 1988. VI, 218 pages. 1990.

Vol. 1435: St. Ruscheweyh, E.B. Saff, L.C. Salinas, R.S. Varga (Eds.), Computational Methods and Function Theory. Proceedings, 1989. VI, 211 pages. 1990.

Vol. 1436: S. Xambó-Descamps (Ed.), Enumerative Geometry. Proceedings, 1987. V, 303 pages. 1990.

Vol. 1437: H. Inassaridze (Ed.), K-theory and Homological Algebra. Seminar, 1987–88. V, 313 pages. 1990.

Vol. 1438: P.G. Lemarié (Ed.) Les Ondelettes en 1989. Seminar. IV, 212 pages. 1990.

Vol. 1439: E. Bujalance, J.J. Etayo, J.M. Gamboa, G. Gromadzki. Automorphism Groups of Compact Bordered Klein Surfaces: A Combinatorial Approach. XIII, 201 pages. 1990.

Vol. 1440: P. Latiolais (Ed.), Topology and Combinatorial Groups Theory. Seminar, 1985–1988. VI, 207 pages. 1990.

Vol. 1441: M. Coornaert, T. Delzant, A. Papadopoulos. Géométrie et théorie des groupes. X, 165 pages. 1990.

Vol. 1442: L. Accardi, M. von Waldenfels (Eds.), Quantum Probability and Applications V. Proceedings, 1988. VI, 413 pages. 1990.

Vol. 1443: K.H. Dovermann, R. Schultz, Equivariant Surgery Theories and Their Periodicity Properties. VI, 227 pages. 1990.

Vol. 1444: H. Korezlioglu, A.S. Ustunel (Eds.), Stochastic Analysis and Related Topics VI. Proceedings, 1988. V, 268 pages. 1990.

Vol. 1445: F. Schulz, Regularity Theory for Quasilinear Elliptic Systems and – Monge Ampère Equations in Two Dimensions. XV, 123 pages. 1990.

Vol. 1446: Methods of Nonconvex Analysis. Seminar, 1989. Editor: A. Cellina. V, 206 pages. 1990.

Vol. 1447: J.-G. Labesse, J. Schwermer (Eds), Cohomology of Arithmetic Groups and Automorphic Forms. Proceedings, 1989. V, 358 pages. 1990.

Vol. 1448: S.K. Jain, S.R. López-Permouth (Eds.), Non-Commutative Ring Theory. Proceedings, 1989. V, 166 pages. 1990.

Vol. 1449: W. Odyniec, G. Lewicki, Minimal Projections in Banach Spaces. VIII, 168 pages. 1990.

Vol. 1450: H. Fujita, T. Ikebe, S.T. Kuroda (Eds.), Functional-Analytic Methods for Partial Differential Equations. Proceedings, 1989. VII, 252 pages. 1990.

Vol. 1451: L. Alvarez-Gaumé, E. Arbarello, C. De Concini, N.J. Hitchin, Global Geometry and Mathematical Physics. Montecatini Terme 1988. Seminar. Editors: M. Francaviglia, F. Gherardelli. IX, 197 pages. 1990.

Vol. 1452: E. Hlawka, R.F. Tichy (Eds.), Number-Theoretic Analysis. Seminar, 1988–89. V, 220 pages. 1990.

Vol. 1453: Yu.G. Borisovich, Yu.E. Gliklikh (Eds.), Global Analysis – Studies and Applications IV. V, 320 pages. 1990.

Vol. 1454: F. Baldassari, S. Bosch, B. Dwork (Eds.), p-adic Analysis. Proceedings, 1989. V, 382 pages. 1990.

Vol. 1455: J.-P. Françoise, R. Roussarie (Eds.), Bifurcations of Planar Vector Fields. Proceedings, 1989. VI, 396 pages. 1990.

Vol. 1456: L.G. Kovács (Ed.), Groups – Canberra 1989. Proceedings. XII, 198 pages. 1990.

Vol. 1457: O. Axelsson, L.Yu. Kolotilina (Eds.), Preconditioned Conjugate Gradient Methods. Proceedings, 1989. V, 196 pages. 1990.

Vol. 1458: R. Schaaf, Global Solution Branches of Two Point Boundary Value Problems. XIX, 141 pages. 1990.

Vol. 1459: D. Tiba, Optimal Control of Nonsmooth Distributed Parameter Systems. VII, 159 pages. 1990.

Vol. 1460: G. Toscani, V. Boffi, S. Rionero (Eds.), Mathematical Aspects of Fluid Plasma Dynamics. Proceedings, 1988. V, 221 pages. 1991.

Vol. 1461: R. Gorenflo, S. Vessella, Abel Integral Equations. VII, 215 pages. 1991.

Vol. 1462: D. Mond, J. Montaldi (Eds.), Singularity Theory and its Applications. Warwick 1989, Part I. VIII, 405 pages. 1991.

Vol. 1463: R. Roberts, I. Stewart (Eds.), Singularity Theory and its Applications. Warwick 1989, Part II. VIII, 322 pages. 1991.

Vol. 1464: D. L. Burkholder, E. Pardoux, A. Sznitman, Ecole d'Eté de Probabilités de Saint- Flour XIX-1989. Editor: P. L. Hennequin. VI, 256 pages. 1991.

Vol. 1465: G. David, Wavelets and Singular Integrals on Curves and Surfaces. X, 107 pages. 1991.

Vol. 1466: W. Banaszczyk, Additive Subgroups of Topological Vector Spaces. VII, 178 pages. 1991.

Vol. 1467: W. M. Schmidt, Diophantine Approximations and Diophantine Equations. VIII, 217 pages. 1991.

Vol. 1468: J. Noguchi, T. Ohsawa (Eds.), Prospects in Complex Geometry. Proceedings, 1989. VII, 421 pages. 1991.

Vol. 1469: J. Lindenstrauss, V. D. Milman (Eds.), Geometric Aspects of Functional Analysis. Seminar 1989-90. XI, 191 pages. 1991.

Vol. 1470: E. Odell, H. Rosenthal (Eds.), Functional Analysis. Proceedings, 1987-89. VII, 199 pages. 1991.

Vol. 1471: A. A. Panchishkin, Non-Archimedean L-Functions of Siegel and Hilbert Modular Forms. VII, 157 pages. 1991.

Vol. 1472: T. T. Nielsen, Bose Algebras: The Complex and Real Wave Representations. V, 132 pages. 1991.

Vol. 1473: Y. Hino, S. Murakami, T. Naito, Functional Differential Equations with Infinite Delay. X, 317 pages. 1991.

Vol. 1474: S. Jackowski, B. Oliver, K. Pawałowski (Eds.), Algebraic Topology, Poznań 1989. Proceedings. VIII, 397 pages. 1991.

Vol. 1475: S. Busenberg, M. Martelli (Eds.), Delay Differential Equations and Dynamical Systems. Proceedings, 1990. VIII, 249 pages. 1991.

Vol. 1476: M. Bekkali, Topics in Set Theory. VII, 120 pages. 1991.

Vol. 1477: R. Jajte, Strong Limit Theorems in Noncommutative $L_2$-Spaces. X, 113 pages. 1991.

Vol. 1478: M.-P. Malliavin (Ed.), Topics in Invariant Theory. Seminar 1989-1990. VI, 272 pages. 1991.

Vol. 1479: S. Bloch, I. Dolgachev, W. Fulton (Eds.), Algebraic Geometry. Proceedings, 1989. VII, 300 pages. 1991.

Vol. 1480: F. Dumortier, R. Roussarie, J. Sotomayor, H. Żoładek, Bifurcations of Planar Vector Fields: Nilpotent Singularities and Abelian Integrals. VIII, 226 pages. 1991.

Vol. 1481: D. Ferus, U. Pinkall, U. Simon, B. Wegner (Eds.), Global Differential Geometry and Global Analysis. Proceedings, 1991. VIII, 283 pages. 1991.

Vol. 1482: J. Chabrowski, The Dirichlet Problem with $L^2$-Boundary Data for Elliptic Linear Equations. VI, 173 pages. 1991.

Vol. 1483: E. Reithmeier, Periodic Solutions of Nonlinear Dynamical Systems. VI, 171 pages. 1991.

Vol. 1484: H. Delfs, Homology of Locally Semialgebraic Spaces. IX, 136 pages. 1991.

Vol. 1485: J. Azéma, P. A. Meyer, M. Yor (Eds.), Séminaire de Probabilités XXV. VIII, 440 pages. 1991.

Vol. 1486: L. Arnold, H. Crauel, J.-P. Eckmann (Eds.), Lyapunov Exponents. Proceedings, 1990. VIII, 365 pages. 1991.

Vol. 1487: E. Freitag, Singular Modular Forms and Theta Relations. VI, 172 pages. 1991.

Vol. 1488: A. Carboni, M. C. Pedicchio, G. Rosolini (Eds.), Category Theory. Proceedings, 1990. VII, 494 pages. 1991.

Vol. 1489: A. Mielke, Hamiltonian and Lagrangian Flows on Center Manifolds. X, 140 pages. 1991.

Vol. 1490: K. Metsch, Linear Spaces with Few Lines. XIII, 196 pages. 1991.

Vol. 1491: E. Lluis-Puebla, J.-L. Loday, H. Gillet, C. Soulé, V. Snaith, Higher Algebraic K-Theory: an overview. IX, 164 pages. 1992.

Vol. 1492: K. R. Wicks, Fractals and Hyperspaces. VIII, 168 pages. 1991.

Vol. 1493: E. Benoît (Ed.), Dynamic Bifurcations. Proceedings, Luminy 1990. VII, 219 pages. 1991.

Vol. 1494: M.-T. Cheng, X.-W. Zhou, D.-G. Deng (Eds.), Harmonic Analysis. Proceedings, 1988. IX, 226 pages. 1991.

Vol. 1495: J. M. Bony, G. Grubb, L. Hörmander, H. Komatsu, J. Sjöstrand, Microlocal Analysis and Applications. Montecatini Terme, 1989. Editors: L. Cattabriga, L. Rodino. VII, 349 pages. 1991.

Vol. 1496: C. Foias, B. Francis, J. W. Helton, H. Kwakernaak, J. B. Pearson, $H_\infty$-Control Theory. Como, 1990. Editors: E. Mosca, L. Pandolfi. VII, 336 pages. 1991.

Vol. 1497: G. T. Herman, A. K. Louis, F. Natterer (Eds.), Mathematical Methods in Tomography. Proceedings 1990. X, 268 pages. 1991.

Vol. 1498: R. Lang, Spectral Theory of Random Schrödinger Operators. X, 125 pages. 1991.

Vol. 1499: K. Taira, Boundary Value Problems and Markov Processes. IX, 132 pages. 1991.

Vol. 1500: J.-P. Serre, Lie Algebras and Lie Groups. VII, 168 pages. 1992.

Vol. 1501: A. De Masi, E. Presutti, Mathematical Methods for Hydrodynamic Limits. IX, 196 pages. 1991.

Vol. 1502: C. Simpson, Asymptotic Behavior of Monodromy. V, 139 pages. 1991.

Vol. 1503: S. Shokranian, The Selberg-Arthur Trace Formula (Lectures by J. Arthur). VII, 97 pages. 1991.

Vol. 1504: J. Cheeger, M. Gromov, C. Okonek, P. Pansu, Geometric Topology: Recent Developments. Editors: P. de Bartolomeis, F. Tricerri. VII, 197 pages. 1991.

Vol. 1505: K. Kajitani, T. Nishitani, The Hyperbolic Cauchy Problem. VII, 168 pages. 1991.

Vol. 1506: A. Buium, Differential Algebraic Groups of Finite Dimension. XV, 145 pages. 1992.

Vol. 1507: K. Hulek, T. Peternell, M. Schneider, F.-O. Schreyer (Eds.), Complex Algebraic Varieties. Proceedings, 1990. VII, 179 pages. 1992.

Vol. 1508: M. Vuorinen (Ed.), Quasiconformal Space Mappings. A Collection of Surveys 1960-1990. IX, 148 pages. 1992.

Vol. 1509: J. Aguadé, M. Castellet, F. R. Cohen (Eds.), Algebraic Topology - Homotopy and Group Cohomology. Proceedings, 1990. X, 330 pages. 1992.

Vol. 1510: P. P. Kulish (Ed.), Quantum Groups. Proceedings, 1990. XII, 398 pages. 1992.

Vol. 1511: B. S. Yadav, D. Singh (Eds.), Functional Analysis and Operator Theory. Proceedings, 1990. VIII, 223 pages. 1992.

Vol. 1512: L. M. Adleman, M.-D. A. Huang, Primality Testing and Abelian Varieties Over Finite Fields. VII, 142 pages. 1992.

Vol. 1513: L. S. Block, W. A. Coppel, Dynamics in One Dimension. VIII, 249 pages. 1992.

Vol. 1514: U. Krengel, K. Richter, V. Warstat (Eds.), Ergodic Theory and Related Topics III, Proceedings, 1990. VIII, 236 pages. 1992.

Vol. 1515: E. Ballico, F. Catanese, C. Ciliberto (Eds.), Classification of Irregular Varieties. Proceedings, 1990. VII, 149 pages. 1992.

Vol. 1516: R. A. Lorentz, Multivariate Birkhoff Interpolation. IX, 192 pages. 1992.

Vol. 1517: K. Keimel, W. Roth, Ordered Cones and Approximation. VI, 134 pages. 1992.

Vol. 1518: H. Stichtenoth, M. A. Tsfasman (Eds.), Coding Theory and Algebraic Geometry. Proceedings, 1991. VIII, 223 pages. 1992.

Vol. 1519: M. W. Short, The Primitive Soluble Permutation Groups of Degree less than 256. IX, 145 pages. 1992.

Vol. 1520: Yu. G. Borisovich, Yu. E. Gliklikh (Eds.), Global Analysis – Studies and Applications V. VII, 284 pages. 1992.

Vol. 1521: S. Busenberg, B. Forte, H. K. Kuiken, Mathematical Modelling of Industrial Process. Bari, 1990. Editors: V. Capasso, A. Fasano. VII, 162 pages. 1992.

Vol. 1522: J.-M. Delort, F. B. I. Transformation. VII, 101 pages. 1992.

Vol. 1523: W. Xue, Rings with Morita Duality. X, 168 pages. 1992.

Vol. 1524: M. Coste, L. Mahé, M.-F. Roy (Eds.), Real Algebraic Geometry. Proceedings, 1991. VIII, 418 pages. 1992.

Vol. 1525: C. Casacuberta, M. Castellet (Eds.), Mathematical Research Today and Tomorrow. VII, 112 pages. 1992.

Vol. 1526: J. Azéma, P. A. Meyer, M. Yor (Eds.), Séminaire de Probabilités XXVI. X, 633 pages. 1992.

Vol. 1527: M. I. Freidlin, J.-F. Le Gall, Ecole d'Eté de Probabilités de Saint-Flour XX – 1990. Editor: P. L. Hennequin. VIII, 244 pages. 1992.

Vol. 1528: G. Isac, Complementarity Problems. VI, 297 pages. 1992.

Vol. 1529: J. van Neerven, The Adjoint of a Semigroup of Linear Operators. X, 195 pages. 1992.

Vol. 1530: J. G. Heywood, K. Masuda, R. Rautmann, S. A. Solonnikov (Eds.), The Navier-Stokes Equations II – Theory and Numerical Methods. IX, 322 pages. 1992.

Vol. 1531: M. Stoer, Design of Survivable Networks. IV, 206 pages. 1992.

Vol. 1532: J. F. Colombeau, Multiplication of Distributions. X, 184 pages. 1992.

Vol. 1533: P. Jipsen, H. Rose, Varieties of Lattices. X, 162 pages. 1992.

Vol. 1534: C. Greither, Cyclic Galois Extensions of Commutative Rings. X, 145 pages. 1992.

Vol. 1535: A. B. Evans, Orthomorphism Graphs of Groups. VIII, 114 pages. 1992.

Vol. 1536: M. K. Kwong, A. Zettl, Norm Inequalities for Derivatives and Differences. VII, 150 pages. 1992.

Vol. 1537: P. Fitzpatrick, M. Martelli, J. Mawhin, R. Nussbaum, Topological Methods for Ordinary Differential Equations. Montecatini Terme, 1991. Editors: M. Furi, P. Zecca. VII, 218 pages. 1993.

Vol. 1538: P.-A. Meyer, Quantum Probability for Probabilists. X, 287 pages. 1993.

Vol. 1539: M. Coornaert, A. Papadopoulos, Symbolic Dynamics and Hyperbolic Groups. VIII, 138 pages. 1993.

Vol. 1540: H. Komatsu (Ed.), Functional Analysis and Related Topics, 1991. Proceedings. XXI, 413 pages. 1993.

Vol. 1541: D. A. Dawson, B. Maisonneuve, J. Spencer, Ecole d' Eté de Probabilités de Saint-Flour XXI - 1991. Editor: P. L. Hennequin. VIII, 356 pages. 1993.

Vol. 1542: J. Fröhlich, Th. Kerler, Quantum Groups, Quantum Categories and Quantum Field Theory. VII, 431 pages. 1993.

Vol. 1543: A. L. Dontchev, T. Zolezzi, Well-Posed Optimization Problems. XII, 421 pages. 1993.

Vol. 1544: M. Schürmann, White Noise on Bialgebras. VII, 146 pages. 1993.

Vol. 1545: J. Morgan, K. O'Grady, Differential Topology of Complex Surfaces. VIII, 224 pages. 1993.

Vol. 1546: V. V. Kalashnikov, V. M. Zolotarev (Eds.), Stability Problems for Stochastic Models. Proceedings, 1991. VIII, 229 pages. 1993.

Vol. 1547: P. Harmand, D. Werner, W. Werner, M-ideals in Banach Spaces and Banach Algebras. VIII, 387 pages. 1993.

Vol. 1548: T. Urabe, Dynkin Graphs and Quadrilateral Singularities. VI, 233 pages. 1993.

Vol. 1549: G. Vainikko, Multidimensional Weakly Singular Integral Equations. XI, 159 pages. 1993.

Vol. 1550: A. A. Gonchar, E. B. Saff (Eds.), Methods of Approximation Theory in Complex Analysis and Mathematical Physics IV, 222 pages, 1993.

Vol. 1551: L. Arkeryd, P. L. Lions, P.A. Markowich, S.R. S. Varadhan. Nonequilibrium Problems in Many-Particle Systems. Montecatini, 1992. Editors: C. Cercignani, M. Pulvirenti. VII, 158 pages 1993.

Vol. 1552: J. Hilgert, K.-H. Neeb, Lie Semigroups and their Applications. XII, 315 pages. 1993.

Vol. 1553: J.-L- Colliot-Thélène, J. Kato, P. Vojta. Arithmetic Algebraic Geometry. Trento, 1991. Editor: E. Ballico. VII, 223 pages. 1993.

Vol. 1554: A. K. Lenstra, H. W. Lenstra, Jr. (Eds.), The Development of the Number Field Sieve. VIII, 131 pages. 1993.

Vol. 1555: O. Liess, Conical Refraction and Higher Microlocalization. X, 389 pages. 1993.

Vol. 1556: S. B. Kuksin, Nearly Integrable Infinite-Dimensional Hamiltonian Systems. XXVII, 101 pages. 1993.

Vol. 1557: J. Azéma, P. A. Meyer, M. Yor (Eds.), Séminaire de Probabilités XXVII. VI, 327 pages. 1993.

Vol. 1558: T. J. Bridges, J. E. Furter, Singularity Theory and Equivariant Symplectic Maps. VI, 226 pages. 1993.

Vol. 1559: V. G. Sprindžuk, Classical Diophantine Equations. XII, 228 pages. 1993.

Vol. 1560: T. Bartsch, Topological Methods for Variational Problems with Symmetries. X, 152 pages. 1993.

Vol. 1561: I. S. Molchanov, Limit Theorems for Unions of Random Closed Sets. X, 157 pages. 1993.

Vol. 1562: G. Harder, Eisensteinkohomologie und die Konstruktion gemischter Motive. XX, 184 pages. 1993.

Vol. 1563: E. Fabes, M. Fukushima, L. Gross, C. Kenig, M. Röckner, D. W. Stroock, Dirichlet Forms. Varenna, 1992. Editors: G. Dell'Antonio, U. Mosco. VII, 245 pages. 1993.

Vol. 1564: J. Jorgenson, S. Lang, Basic Analysis of Regularized Series and Products. IX, 122 pages. 1993.

Vol. 1565: L. Boutet de Monvel, C. De Concini, C. Procesi, P. Schapira, M. Vergne. D-modules, Representation Theory, and Quantum Groups. Venezia, 1992. Editors: G. Zampieri, A. D'Agnolo. VII, 217 pages. 1993.

Vol. 1566: B. Edixhoven, J.-H. Evertse (Eds.), Diophantine Approximation and Abelian Varieties. XIII, 127 pages. 1993.

Vol. 1567: R. L. Dobrushin, S. Kusuoka, Statistical Mechanics and Fractals. VII, 98 pages. 1993.

Vol. 1568: F. Weisz, Martingale Hardy Spaces and their Application in Fourier Analysis. VIII, 217 pages. 1994.

Vol. 1569: V. Totik, Weighted Approximation with Varying Weight. VI, 117 pages. 1994.

Vol. 1570: R. deLaubenfels, Existence Families, Functional Calculi and Evolution Equations. XV, 234 pages. 1994.

Vol. 1571: S. Yu. Pilyugin, The Space of Dynamical Systems with the $C^0$-Topology. X, 188 pages. 1994.

Vol. 1572: L. Göttsche, Hilbert Schemes of Zero-Dimensional Subschemes of Smooth Varieties. IX, 196 pages. 1994.

Vol. 1573: V. P. Havin, N. K. Nikolski (Eds.), Linear and Complex Analysis – Problem Book 3 – Part I. XXII, 489 pages. 1994.

Vol. 1574: V. P. Havin, N. K. Nikolski (Eds.), Linear and Complex Analysis – Problem Book 3 – Part II. XXII, 507 pages. 1994.

Vol. 1575: M. Mitrea, Clifford Wavelets, Singular Integrals, and Hardy Spaces. XI, 116 pages. 1994.

Vol. 1576: K. Kitahara, Spaces of Approximating Functions with Haar-Like Conditions. X, 110 pages. 1994.

Vol. 1577: N. Obata, White Noise Calculus and Fock Space. X, 183 pages. 1994.

Vol. 1578: J. Bernstein, V. Lunts, Equivariant Sheaves and Functors. V, 139 pages. 1994.

Vol. 1579: N. Kazamaki, Continuous Exponential Martingales and *BMO*. VII, 91 pages. 1994.

Vol. 1580: M. Milman, Extrapolation and Optimal Decompositions with Applications to Analysis. XI, 161 pages. 1994.

Vol. 1581: D. Bakry, R. D. Gill, S. A. Molchanov, Lectures on Probability Theory. Editor: P. Bernard. VIII, 420 pages. 1994.

Vol. 1582: W. Balser, From Divergent Power Series to Analytic Functions. X, 108 pages. 1994.

Vol. 1583: J. Azéma, P. A. Meyer, M. Yor (Eds.), Séminaire de Probabilités XXVIII. VI, 334 pages. 1994.

Vol. 1584: M. Brokate, N. Kenmochi, I. Müller, J. F. Rodriguez, C. Verdi, Phase Transitions and Hysteresis. Montecatini Terme, 1993. Editor: A. Visintin. VII, 291 pages. 1994.

Vol. 1585: G. Frey (Ed.), On Artin's Conjecture for Odd 2-dimensional Representations. VIII, 148 pages. 1994.

Vol. 1586: R. Nillsen, Difference Spaces and Invariant Linear Forms. XII, 186 pages. 1994.

Vol. 1587: N. Xi, Representations of Affine Hecke Algebras. VIII, 137 pages. 1994.

Vol. 1588: C. Scheiderer, Real and Étale Cohomology. XXIV, 273 pages. 1994.

Vol. 1589: J. Bellissard, M. Degli Esposti, G. Forni, S. Graffi, S. Isola, J. N. Mather, Transition to Chaos in Classical and Quantum Mechanics. Montecatini, 1991. Editor: S. Graffi. VII, 192 pages. 1994.

Vol. 1590: P. M. Soardi, Potential Theory on Infinite Networks. VIII, 187 pages. 1994.

Vol. 1591: M. Abate, G. Patrizio, Finsler Metrics – A Global Approach. IX, 180 pages. 1994.

Vol. 1592: K. W. Breitung, Asymptotic Approximations for Probability Integrals. IX, 146 pages. 1994.

Vol. 1593: D. Goldfeld, J. Jorgenson, S. Lang, Explicit Formulas for Regularized Products and Series. VIII, 154 pages. 1994.

Vol. 1594: M. Green, J. Murre, C. Voisin, Algebraic Cycles and Hodge Theory. Torino, 1993. Editors: A. Albano, F. Bardelli. VII, 275 pages. 1994.

Vol. 1595: R.D.M. Accola, Topics in the Theory of Riemann Surfaces. IX, 105 pages. 1994.

Vol. 1596: L. Heindorf, L. B. Shapiro, Nearly Projective Boolean Algebras. X, 202 pages. 1994.